LIFESPAN

LIFESPAN

HOLLY ZIMMERMANN

IngramSpark

Disclaimer: *Lifespan* is a work of fiction. All persons, events and occurrences are imaginary. Any similarities to people or factual events are by chance and based on personal ideas and experiences of the author. Any medical or mathematical mistakes are unintentional and solely the responsibility of the author.

Copyright © 2021 by Holly Zimmermann

No part of this book may be reproduced or used by any means without prior written permission of the author, with the exception of brief quotations for articles and reviews.

For permission requests, please email the author at:
hollyzimmermann@yahoo.com

ISBN: 978-1-7365780-0-1 (paperback)
ISBN: 978-1-7365780-1-8 (Ebook)

Cover and interior design: Holly Zimmermann

Other works by Holly Zimmermann:

Ultramarathon Mom: From the Sahara to the Arctic
Ultramarathon Mom describes how Holly, mother of four, has endeavored to take on some of the world's most difficult and dangerous foot races. A grueling 160-mile ultramarathon through the Sahara Desert is the core of her story. Tales of running under a scorching sun, living on granola bars and nuts, and sleeping on the ground of an open tent are balanced with heartwarming stories of friendship and camaraderie. Interspersed between her Sahara adventures, Holly recalls previous races and training runs full of mishaps, written in her own humorous style. After having conquered the Sahara Desert, she takes on the Polar Circle Marathon in Greenland. There she experienced temperatures cold enough to cause frostbite within minutes. With chains on her running shoes and four layers of clothing, Holly describes how she ran in one of the most harsh but beautiful places on earth. *Ultramarathon Mom: From the Sahara to the Arctic* tells a unique story and delivers a meaningful message: Live your dreams.

Running Everest: Adventures at the Top of the World
Running Everest tells the story of a group of adventurers from around the globe who embark on a remarkable journey through the Khumbu Valley of Nepal, battling high-altitude sickness, deplorable sanitary conditions, freezing temperatures...and enjoying every minute of it! When they reach their destination, Mount Everest Base Camp, they turn around and run a marathon, the highest marathon in the world, back to civilization. Are they extremists? Or the new generation of ordinary people? Written with humor and passion, *Running Everest* explores the culture, inhabitants, and the delicate balance of Hinduism and Buddhism in the breathtaking Himalayas, topped off by an exhilarating race over glacial moraines, high altitude plateaus, and steep rocky climbs, all in the shadow of the highest mountain on earth.

Chapter One

Professor Eduardo Rodriguez shut down his computer and breathed a sigh of relief. Everything had been taken care of, his holdings liquidated, cash transferred among the accounts of his wife and children, and trust funds arranged for his grandchildren. All of the instructions had been clearly stated in his last will and testament, which had been finalized with his attorney in Sao Paulo that morning.

He closed his office door to avoid being interrupted, then picked up the phone to make the first of two calls.

Miguel was at his desk when he received his father's call. "Hey, Dad", then a brief pause, "everything alright?" Eduardo smiled as he heard the familiar voice that answered. His son, Miguel, lived in Brasilia, where he'd moved shortly after college to work as an architect in a small firm.

"Sure, son, everything is fine. I just wanted to see how you're doing." Eduardo didn't often call his son at the office, it was usually evenings or weekends that they talked, but this time it couldn't be avoided.

Miguel leaned back in his chair and let out a sigh. "I'm doing well. We've got a lot of new projects here, and some of it is really interesting. I'm enjoying it, but of course the downside is that it keeps me away from Katie and the kids." He spun his chair around to look out the window next to him. "Oh! Speaking of the kids, I have some news for you."

"Really? What is it?" Hearing about the lives of his children and grandchildren was one of the most simple but greatest pleasures in Eduardo's life. Whatever the topic, he was always ready to listen and made it a priority to stay in touch, despite the distance and their busy schedules.

"Katie is pregnant again. We're expecting in September."

"That's wonderful!" Eduardo was overcome with emotion. "I'm really happy to hear that. I know you two wanted a large family." He glanced over at the photo of his son and family, which stood on his desk. "Four children, a lovely wife, a successful career, you really have done well for yourself. I'm very proud of you." Eduardo had always made an effort to speak openly with his family.

"Thanks, Dad. Yeah, honestly I'm really happy with my life at the moment. Don't know if I could have planned it any better." He looked at the stack of work in front of him, begging for his attention. "But I'm pretty much at my limit right now; don't think I could handle any major catastrophes at the moment," he laughed.

Eduardo felt a sudden pang of guilt, but he knew he couldn't change fate. "Son, we aren't burdened with more than we can bear. Whatever comes your way, you'll be able to handle."

"I know, Dad, you've told me that many times," he said with a smile, knowing how sentimental his father sometimes got.

Eduardo could not hide the emotion in his voice, but it was important for him to express just how he felt. "Miguel, it's true, I am proud of you, and I know that I haven't said it in a long time and certainly not often enough, but I love you very much."

Miguel smiled and took advantage of the rare occasion to reciprocate, and then after a brief mutual silence he quickly changed the subject. "So, have you been watching the World Cup qualifiers?"

"I sure have, son." They went on for a while with predictions for the upcoming games before Miguel was called to a meeting.

After ending the call with his son and collecting his thoughts, his daughter Maria was the next on his list. With her he also spoke similar words of pride and love. It was difficult to end the conversation and he held the phone to his ear long after she had said goodbye.

Eduardo took a look around the office where he'd worked for the majority of his professional life. As his eyes scanned the bookshelves, he was flooded with memories. Each book and each binder told a story. But he knew that right now he had no time to reminisce. He had to keep his priorities straight. So he packed up his things and drove home.

Rosalinda Rodriguez was preparing supper when Eduardo arrived. She was humming a tune as he walked into the kitchen and he wrapped his arms around her from behind. She jumped and let out a gasp. He had caught her by surprise, since she wasn't expecting him for about an hour.

He kissed the nape of her neck and gently caressed her hair. Rosalinda relented and closed her eyes for a moment before questioning what she'd done to deserve such a wonderful husband.

"You've done nothing but be yourself and that's all I've ever asked for. I simply decided to come home early and spend a relaxing evening with you. No crime in that, is there?"

"Certainly not, darling." She went to the sink and rinsed the flour from her hands. "That's wonderful, and the perfect night for it since we're having your favorite meal: seafood empanadas!"

"Oh, Rosa, that's perfect. I love you."

"Me? Or my seafood empanadas?"

"Both," he cleverly responded while stealing a spoonful of seafood from the pan.

"You're in a good mood today," she commented while continuing to bustle about in the kitchen. "I talked to Miguel. He said you got emotional after hearing his news." Rosa had also learned that day that they would soon be grandparents for the fourth time. Miguel had given her a quick call between meetings, which he often did. "Maybe I should talk to Maria about getting started on her family if it makes you so affectionate," she said with a grin.

"I'm just happy with the way our lives have turned out," Eduardo said. "Our children are stable and happy. We've always had everything that we needed." He laughed, "Well, except for those few years when I was finishing my studies and we were living off my student's stipend." He smiled. "That was a wonderful time though for its own reasons. At least now looking back at it," he shrugged. "And in the end," he hesitated briefly, "well, let's just say I couldn't have asked for more from my life."

"In the end...in the end...," she said. "Why do you say such things?"

They sat together eating supper by candlelight at the long oak table where they had spent countless hours with their family. Eduardo remembered trying to feed Miguel at that same table when he was just a baby; by the end of the meal, both father and son were covered with food. Then along came Maria and she stole his heart, a complete angel, extremely shy with strangers but open and curious with her family. She had such big, beautiful dark eyes as a baby that Eduardo wondered if they were too large for her delicate face. But as she matured, they became her most endearing feature. Like her mother, she was shy, kind and loving.

They reminisced about neighbors and friends who'd spent many hours with them at that table over the years, being spoiled by Rosa's excellent cooking, while drinking wine and talking into the night. And now, although the two of them were sitting alone, they were enjoying one another's company just as much as they always had for the past thirty-eight years.

It began getting late and Rosa cleared the dishes while Eduardo stepped out onto the veranda for some fresh air. The sky was clear and bright as the moon was nearly full. He looked at the constellations of the stars, many he could name, though some he'd forgotten. He breathed in deeply and closed his eyes, hoping to imprint that moment on his mind forever.

Rosa called out to say that she was heading to bed. He decided to follow but only after taking one last look at the beautiful night sky. Then he turned to go inside. He walked through the house, stepping for a brief moment in each room, just to look, just to be sure that everything was orderly, that he'd forgotten nothing. In his dressing room he changed into fresh pajamas then joined his wife under the large woolen blanket.

She was almost asleep when he lay down next to her, but still he leaned over, and whispered, "I love you, Rosa." He waited and saw her lips move with a faint response, "I love you too, Eduardo."

Then he stroked her hair and kissed her lightly on the cheek. Leaning back on his pillow, he closed his eyes and soon fell into a deep, tranquil, and *permanent* sleep.

Chapter Two

Ron Stanley woke before the alarm and wrapped his arms around his wife lying next to him as the sunlight began to creep through the blinds. After a few minutes, she kissed his hand, then slipped out from his hold. He reached over to catch her, but she was gone, already out of bed, into her slippers and across the room. He watched her leave and knew that she was going to their four-year-old Sophie's room, to rouse her out of her sleep. A few minutes later he heard the shower turn on.

He realized then that if he were to lay there much longer he'd drift back to sleep, so he forced himself up and stretched as he sat on the edge of the bed. He looked in the mirror and ran his fingers through his wavy brown hair. Time for another haircut.

Ron Stanley was not someone who stood out in a crowd. Brown hair, brown eyes, medium build and height. He wasn't handsome, at least he didn't think so, although his wife, Caroline, insisted that 'he was nice to look at'. But he was content with his life, and most of the time quite happy. He pulled on a sweatshirt and opened up the blinds to enjoy the morning sun. It looked like it was going to be a gorgeous spring day. March is usually unpredictable in New England, "In like a lion, out like a lamb", as the saying goes. But today was expected to be unseasonably warm, approaching 70F, and continuing throughout the weekend.

He went downstairs where Max was waiting to greet him with a wagging tail. After giving the dog a playful pat and letting him out into the backyard, he put on a pot of coffee. Then he opened the kitchen window and heard the whistling of the paperboy who was just tossing the morning edition onto the front drive.

As he went out to pick up the paper, he noticed his neighbor, old Mrs. Eldridge, who had also just snuck out to grab her own newspaper; she was trying to hurry back inside before anyone saw her in her bathrobe. Ron couldn't resist the urge and called out to her with an enthusiastic, "Good morning, Mrs. Eldridge!" She was startled by the greeting and offered a quick wave as she scurried back through her perfectly manicured gardens and into her Cape Cod style home.

Ron sat down at the kitchen table with a glass of orange juice and looked over the headlines. The Red Sox finally won a game after an eight game losing streak. Four to three against the Yanks! Well, hallelujah!

Then Sophie sauntered into the room wearing a pretty pink sun-dress. "Good morning, daddy," she said with a huge grin.

"Good morning, honey. Wow, don't you look beautiful," Ron said with all honesty, but thinking that the dress should have stayed in her closet until at least June. He decided not to dampen his daughter's spirits so he simply added, "You might want to take a sweater with you today. Just in case."

"Okay, daddy," she said as she bent down to pick up her stuffed *Fuzzy Bear* and give him a good morning kiss.

He watched his daughter skip to the back door to let their dog in, which was followed by the two of them cuddling and Sophie practicing giving him dog-trick commands. Max was an eight-year-old Golden Retriever, and was up for anything when it came to Sophie. He had a heart as good as gold, and Ron's wife was sure that was how this breed got their name.

Ron filled his daughter's cereal bowl with Cheerios and set it at her place at the table just as she slid into her chair.

Caroline entered the kitchen and Ron poured two cups of coffee. She kissed him quickly on the lips and he handed her a mug from which she immediately took a gulp. "Mmmm. Now I'm awake."

She began collecting a few papers from the countertop and packed them into her briefcase. "Oh, good, Sophie already has her breakfast. I'll drop her off this morning." Another gulp of coffee and her cup was empty. "I've got to go in early anyway since I need to get out by four o'clock in order to make my doctor's appointment."

"All right," Ron said. "I'd better get in the shower now, or I'm going to be late for work." He gave his wife and daughter each a kiss, then as he was leaving the room he added almost as an afterthought, "Hey, do you want me to meet you at the doctor?"

Caroline had not been feeling well for the past couple of months. There had been nothing to send up alarm bells, just more tired than usual and some muscle aches. Their first suspicion was that maybe she was pregnant, but an over-the-counter test unfortunately proved that theory wrong. With nothing but the occasional dose of Tylenol, the symptoms hadn't gone away like a cold usually does after a week or two. They figured it was a virus. An annoying, persistent one.

She'd finally gone to see her general practitioner the week before, after her mother stepped in and urged her to do so. She underwent a full examination, got a prescription for some high potency vitamins, and the doctor took urine and blood samples.

The results would be back today.

"No, I'll be fine. I'm sure the tests will show it's just a viral infection or something that I just haven't kicked. Or maybe it's bacterial and she'll give me some antibiotics and send me on my way."

Ron watched her pull a loaf of bread from the cupboard and pop a couple of slices into the toaster. "I'm sure you're right," he said. "But if you change your mind give me a call at the office, or actually my cell since I'll probably be out at the site most of the day."

"You're lucky daddy, you get to be outside all day! I want to play outside too!"

He knelt down next to his daughter. "I'd love to be able to play outside, but I'm going to be working. The construction site is no place for play. It's dirty and full of large trucks and building materials, not like a park or our backyard. But I'll promise you something," Ron said. "Tonight we'll have dinner outside on the deck, the first barbecue of the year. How does that sound?"

"Yippee!" Sophie turned to her mom for approval.

"I guess it's settled then." Caroline happily agreed. "I'll pick up some burgers on my way home from the doctor."

Sophie sat there with a wide grin on her face, clearly looking forward to coming home that afternoon and spending time with her family.

The three of them had that in common.

Ron winked at his daughter, gave each of his girls another kiss, and then slipped out of the room.

After a long day, Ron pulled into the daycare at 6 p.m. and immediately saw Sophie running up to greet him with the same wide smile she'd been wearing that morning.

"Daddy!"

Ron scooped his daughter up in his arms. "Hi sweetheart! Did you have a nice day?"

"Yep, it was fun, but the best is tonight, we're barbecuing, remember?"

"Tonight? Are you sure? At our house?"

"Yes daddy, we set it all up at breakfast. Mom's buying hamburgers!"

"Well, if you're sure, then it must be true. Let's go inside, get your things and say goodbye, okay?"

In the car they talked about Sophie's day and the art project that she was working on. She was in the process of building a paper maché horse and planned to paint it pink, her favorite color.

"But horses aren't pink, Sophie." Ron tried to give a little parental guidance.

Sophie quickly straightened him out by adding, "But daddy, they aren't two feet tall either. It's a pretend horse."

He feared it wouldn't be long before she was outsmarting him on more than just pink horses.

They pulled into the driveway and saw that Caroline was

already home; her car was in the garage. Sophie jumped out and ran inside the house while Ron walked out to the mailbox to pick up the mail. Mrs. Eldridge was next door planting bulbs in her garden.

"Beautiful day isn't it, Mrs. Eldridge?" he said.

"Yes, it is Mr. Stanley, nicest so far this year. Perhaps we'll have an early summer." She immediately turned back to her work, as though her garden was the only company that she truly wanted.

Ron entered the house and could hear Sophie in the living room. She was talking in the sweet voice that she used with her dolls, so he assumed that she'd pulled out Barbie to join the barbecue. Then he heard Caroline's voice from the same room, although it didn't sound quite like her, it was almost as if she had the flu and was congested. He walked toward the direction of the voices and heard Sophie say, "It's okay, mommy, don't cry, we're having a barbecue tonight."

Why is Caroline crying, he wondered? Maybe she is just playing along with one of Sophie's games? But then he remembered the doctor's appointment.

He rushed into the living room where he found Caroline curled up on the couch, eyes red and puffy. She was trying to smile as she held Sophie's hand and spoke to her gently. When she looked up at him he knew immediately that he was about to receive some bad news.

"Sophie, why don't you go put on some jeans and a sweater. It's going to be a little chilly outside tonight." Ron could barely get those words out but he knew that he needed to talk to Caroline alone.

"Okay, daddy," she said as she skipped out of the room.

Ron sat down next to Caroline on the couch but didn't know where to begin, he lifted her up and hugged her as tight as he could. She was almost limp in his arms, completely drained of all energy.

"What did the doctor say?"

She looked down at the tissue in her wringing hands and began to cry.

"Come on honey, tell me." He lifted her chin so that their eyes met.

"It...it's...bad..." She burst into more tears.

"Talk to me. What did she say?"

She blew her nose, wiped her eyes and began to talk. "Well, initially Doctor Mansfield just told me that she had gotten the results of the blood test and that she was concerned. She...well, she noticed that some markers were high, but she didn't want to worry me until she had some definite answers, so she took more blood samples to send off for further tests—"

"Okay, honey, that doesn't sound like such bad news—"

"I'm not finished yet, Ron."

"Oh, go on." He felt his chest tighten.

"She asked me lots of questions. Like about my digestion, headaches, swollen glands and if I go to the gynecologist regularly for screening. She said everything in that respect sounded good..."

"But...?" Ron knew something was coming.

"Then she took a bone marrow biopsy."

"What? Bone marrow?" Ron felt like his head was spinning. "Right there in her office? Isn't that a complicated procedure?"

"That's what I thought too," Caroline continued after blowing her nose, "but it was fairly simple. She took a sample from my hip. With a local anesthetic, a long needle, and, well...in a few minutes it was all over."

"But, I still don't understand. What is she looking for? When will we have the results?"

A tear fell from Caroline's eye, then another. "Ron, I have the results already, at least the preliminary ones, she'll send the samples off to the lab to get confirmation." She took a deep breath, then clarified everything for her husband. "All she had to do was look at the sample under the microscope. It confirmed her suspicions."

"Which were?" he asked.

"Ron, I have leukemia." Once she said the words she seemed to strengthen herself. She sat up and wiped the tears from her face.

"Is she sure? How can that be? You're healthy."

"I feel healthy," Caroline said, "but according to Doctor Mansfield that will probably change soon."

Ron was speechless. He wasn't sure what to do or say, so he simply pulled her close to him again and rocked her gently in his arms. He realized the conversation was far from over but they couldn't talk openly with Sophie in the next room.

"Listen," Ron said, "I'll call my parents and ask them to come get Sophie. They'll take her to McDonald's or something. Then we can talk."

"No, she is so excited about our barbecue." Caroline stood up and took a deep breath. "We'll talk later, after we put her to bed."

Ron was caught between concern for his wife and for his

daughter. "Are you sure? I'm positive that it would be no problem for my parents."

"Yes, I'm sure. Now let's try as best we can to act as though everything is fine. We'll have a nice dinner and try to enjoy the evening as much as possible. Sophie is so excited and that will hopefully cheer me up too."

Sophie was humming a tune as she descended the stairs. She then quieted down and peeked around the corner. A moment later she jumped into the living room, "Boo!" she shouted.

Ron and Caroline both laughed.

Sometimes laughter is the best medicine.

"So," Ron said as enthusiastically as he could, "who's ready for burgers?"

Later that evening, after Ron had put Sophie to bed, he came back into the living room to find Caroline lying on the couch. She lifted her head when she heard him, but immediately laid it back down as he sat next to her.

"How do you feel?"

"I'm exhausted."

"Do you want to go to bed?"

Caroline forced herself to sit up. "No, not yet. We need to talk."

"But if you're not feeling up to it—"

"Ron, we need to talk."

"Okay." Ron laid a blanket over his wife's legs. "You know, I was thinking during dinner that I don't even know what leukemia is. I mean, of course I've heard the name a million times and I think it's a type of cancer, but I don't even know for sure."

"Yes, it's cancer. And a nasty one at that."

"What do you mean? Aren't they all?"

"Yes, I suppose they all are." Caroline said. "But what Doctor Mansfield told me today makes me wonder if I have a chance at all."

"Wh...what? What do you mean? Of course you have a chance. You can get chemotherapy or radiation right?"

"Yes, I guess. I...I don't know."

"Did Doctor Mansfield talk about treatment with you?"

"A bit." Caroline said. "But she said that you and I should come back together and she would explain everything in more detail once she got the lab results back. I made an appointment for us."

"But, I thought she was a general practitioner. Is she actually qualified to diagnose you with leukemia?"

"Well, apparently she lost a family member to leukemia recently."

"But that doesn't qualify her to—"

"Yes, I know. But she said something about having continued her studies in Oncology, and that she spends her weekends as a Resident at Rhode Island Hospital. She doesn't rest."

"So, she's planning on switching from being a GP to a specialist?"

Caroline sat up. "I don't know. But I don't think she would have diagnosed me if she wasn't totally sure; we can ask her that when we see her. I did ask her one thing though..."

"Yes, what?"

Caroline was wringing a tissue through her fingers again. "I asked her what the survival rate was for this."

Ron thought his heart skipped a beat. "And? What did she say?"

Caroline didn't answer. She just buried her face in her hands and began to cry again.

Chapter Three

"Jose, as we discussed in class today, the area of a plane region is equal to the integral of its cross-sectional length." Professor Montagna picked up a pencil and began sketching.

"So, if you have a and b which are the max and min x-coordinates of the region, then you must find a line segment, say $c(x)$, which is the length of the intersection with the given region of the vertical line through (x, 0). See?"

Professor Ricardo Montagna drew a hypothetical region and positioned it in the positive xy plane. He then proceeded to draw the vertical line, $c(x)$, through the region and finally the corresponding equation to determine the area.

Jose scratched his head. "Well, what if you only know a and b on the y-axis?"

"You tell me, Jose." He looked at the clock.

Professor Montagne was now giving essentially the same lecture that he had given that morning, only this time it was a review to a single student in his office, who received only average marks at that.

"Can I then use a cross-sectional length along the y-axis?"

"Yes, Jose, that would then be your $c(y)$, and then you integrate over the y-axis coordinates," Professor Montagne pointed at the coordinates on the paper with his pencil and then took a sip of coffee.

"Cool, now I think I get it! Thanks for your help Professor." Jose packed up his books and tucked his pencil behind his ear, then he quickly slipped out of the office.

Professor Montagne breathed a sigh of relief at the student's quick disappearance. He was getting tired of teaching the same subjects year after year. The students' faces change but the questions and difficulties always remain the same. After 20 years of teaching, he had essentially perfected his skills to adapt to the level of the student, to say exactly what was needed in order to help them understand. Now he was just going through the motions. The joy that he experienced as a fresh, young professor had left him somewhere along the line.

He turned back to the work on his desk and found where he had left off before he was interrupted. The notes were well advanced but incomplete; nevertheless, he thought that *this time* he was really going to produce something outstanding.

His new work was a postulate which refuted the natural law of growth and decay and its application to the world population. His brilliance had been to account for the effects of medical, technological and psychological advances as a variable rather than a constant; therefore, it would change over time. Unfortunately, he couldn't exactly formulate this variable with given statistics and he was still struggling with his numbers.

Professor Montagne looked out of his office window over the university quadrangle and could see many students sitting in the sun, some with books, but most just chatting with friends, tossing a Frisbee, or working on their tans. It was Friday afternoon and it looked as though the entire student population had already declared it to be the weekend.

Focusing on his work was getting difficult. Maybe those students weren't so ignorant after all.

He followed their lead and called it a day.

After carefully gathering his files and binding them together, he locked them into his cabinet for safe keeping over the weekend. His tweed jacket hung on a peg behind the door, but before he reached for it he turned again to watch the students on the quad, remembering back to his own days of studying at the university.

Life was so much simpler then, no responsibilities other than his studies, plenty of friends and parties, and even a girlfriend for several years during that time. He had always thought that he and Lucia would marry. But he stayed on at the university to pursue his doctorate and she found a job in a neighboring city. They slowly grew apart until she announced that she had met someone new.

At the time he wasn't upset, thinking that he was still young and there would be many chances to meet new girls, but time went on and no one else entered his life.

Even today he looked back with some regret at having let Lucia go so easily. Maybe if he had put more effort into the relationship instead of his studies then they would now be married with children; but then again he might not be where he is today, a respected research professor at the university.

He switched off the lights, locked the door and turned around just in time to see Professor Amar Kumar, the Head of the Mathematics Department, walk by.

"Calling it an early weekend, Ricardo?"

"Yes, yes, after all the rain we've had, I'd like to get out and enjoy some of this welcome sunshine! Besides my office hours ended ten minutes ago and I certainly don't expect any more students to come talk to me about math when they can be outside."

"I suppose you're right, Ricardo. You, along with Professor Rodriguez."

"What about Professor Rodriguez?" Ricardo asked. "Now that you mention it, I didn't see him at all today. But he didn't say anything about taking a day off when I saw him yesterday."

"Didn't mention it to me either, and I'm his boss! Had to cancel two of his classes. Students didn't seem to mind though!"

"He must have woken up this morning and was lured away by the fine weather," said Ricardo, though he didn't believe that for a moment.

"Yes, you're right," Amar said. "That must have been it. I suppose he just took a well-deserved day off!"

Humph. *A well-deserved day off*? So why should he feel guilty about leaving a little early?

"Well, have a nice weekend, Amar."

"Same for yourself, Ricardo, and try to get some sleep, you're looking a little blanched, if I may say so."

Blanched?

"Yes, perhaps I'm in need of a holiday myself," Ricardo said.

"Spring break is nearly upon us Professor!" Amar said as he strode off down the hall.

Yes, yes, nearly upon us. Ricardo was tired of taking jabs

from Amar, but he was his boss so he tolerated them as best as he could.

Amar Kumar was of Indian descent, though born and raised in Sao Paulo. He was promoted at a relatively early age to the post of Department Head, but most people accredited that advancement to diversification efforts by the University rather than to intelligence and hard work. Nonetheless, the man was amiable and competent, which was why he still held the position a decade later.

Although, sometimes Ricardo got the feeling that Amar was getting restless.

Perhaps a change was in the wind.

On his way out, Ricardo passed by Professor Rodriguez' office. The lights were off and the door closed.

Well, it looked as though our esteemed Professor Rodriguez had indeed extended his weekend. I suppose you can choose your hours when you are such a "highly regarded mathematician", he thought, recalling the quote describing the professor in the university catalogue.

Although Ricardo was well respected in his field, he had always been overshadowed at the university by Professor Rodriguez. To his credit though, his colleague was a genius when it came to numbers, but that only helped to minimize the recognition that he himself had been able to achieve with his own publications.

Ricardo smiled to himself, thinking that everything was about to change. His new postulate would catapult his reputation in the field. He was sure of it.

He just needed to tweak his final variable and then it would be ready to send to the Journal of Mathematical Logic to be made public, at which point his name would replace that of Professor Rodriguez as the world famous mathematician at the University.

Or so he hoped.

As Ricardo walked along the edge of the quad, on the way towards the teacher's parking lot, he recognized several of his students enjoying the weather. Most didn't take notice of him as they were busy with their friends. Those that did see him pass by gave him a hesitant wave or stiffened-up as they tried to force a friendly greeting.

"Have a nice weekend Professor," one of his students called out.

He pretended not to have heard.

It was common knowledge that he was not one of the most well-liked professors on campus. He was known for his tough grading policies and strict attendance requirements. He was a firm believer that students who do not show the initiative to not only be present in class, but to participate as well, should not be rewarded with good grades. He felt that all of his fellow colleagues should use more discipline when it came to teaching, and that subsequent higher standards of expectation would strengthen the reputation of the university.

Driving through the tree-lined campus streets and onto the west-bound thoroughfare leading to his home, he thought about his postulate and what he was missing. He turned it over and over again in his mind.

What was it that he could have possibly overlooked?

He knew that if he discussed it with Professor Rodriguez, it would likely be easily solved, but he wanted to do this on his own. He wanted sole recognition and, perhaps more importantly, he wanted respect from Professor Rodriguez, something which he felt that he had never received from the man.

Maybe he could gather some feedback by discussing it with Ramon Santos, a student who was working on his PhD dissertation with Professor Rodriguez and who had shown extraordinary potential.

Ramon could possibly provide Ricardo with some ideas that would lead him in the right direction and maybe even bring the topic to Rodriguez. That way, Ricardo could receive feedback from the Professor indirectly, without soliciting it.

He turned into his driveway, then parked the car and left it idling while he got out to open the garage. Once inside the house he took off his shoes and jacket and walked into the study which overlooked his small backyard. He went to the sideboard and poured himself a drink, then sat down in his recliner and let out an exhaustive sigh.

He reached into his pants pocket for his pills and swallowed one with a gulp.

Ricardo had suffered from minor seizures since he was a child, and they gradually became more intense as he aged. He hadn't had one in several years, thanks to the new medication, but the tradeoff was that he was forced to take a pill every six hours.

He leaned over and turned on the television to watch the early news. The weather forecast had changed and was now supposed to take a downturn rather suddenly over the weekend, with expected showers lasting through Monday.

No matter, he thought, realizing that he would probably work on his postulate over the weekend anyway. Besides, maybe the bad weather would keep his students in the library studying rather than on the quad socializing!

The onslaught of hunger pangs urged him out of his chair to see what his housekeeper had left for him in the refrigerator for supper.

As he stepped into the kitchen he heard the voice of the news commentator, "In other news, a respected professor at the University of Sao Paulo died last night, the apparent victim of a massive heart attack during his sleep."

Ricardo rushed back into the study just as a photo of Professor Rodriguez was being displayed under the heading, "World Renowned Mathematician dead at the age of 58".

Chapter Four

Inspector Nick da Silva wasn't in a hurry to get into work on Tuesday morning. Although the weekend weather had not been great, he had taken Monday off and was just not ready to face another grueling week at work. He sat and ate a full breakfast with his wife and two boys and perused the sports page of the daily paper.

Nick was in his early thirties, with medium brown hair, tall and strong. He was an avid sports fan and had played soccer in school, but since he was built more like a linebacker than a forward, his agility, or lack thereof, ended his ball playing days early.

His boys were excited because Brazil had beaten Mexico the day before to qualify for the World Cup Championships the following year. He read the article aloud to them about how Brazil had scored the winning goal in the 87th minute to secure the lead and the eventual win. His sons were begging to go see one of the games, but since they were to be played in Germany, Nick told them not to count on it, but that they could definitely go watch them at a public viewing. That seemed to make them happy for the time being.

The normal 20-minute commute from his home to the 8th District Police Headquarters in Sao Paulo took almost 45 minutes. Apparently, the entire city had the same notion to come in late for work that day.

Nick shuffled through the radio stations as he sat in a line of traffic at a red light through three intervals before he finally made it across the intersection. He was relieved when at last he pulled his car into the underground garage at the station. As he turned into his parking space he smiled when he saw the label plate displaying his name bolted into the cement pylon in front of his car.

Along with his promotion to Homicide Inspector came the assigned parking space which, as far as Nick was concerned, was the best perk of the job. Prior to that, since parking in the city was not an option due to over-congestion and vandals, he had been forced to take the train in to work. He hated riding the train. Actually, he despised all forms of public transportation because he wasn't in control of his own motion.

When he was a kid, one of his father's friends had a small plane, a twin engine Cessna, that seated four. He had begged his father to convince his friend to take them for a flight. After months of incessant cajoling, his father finally told him he would be getting his wish. It was just the three of them and Nick was allowed to sit in the front as his view would have been limited from behind. They took off low over the countryside and Nick was awestruck by how small everything looked from above. He could see houses and trees and even the lakes but he couldn't make sense of it all from the new perspective. It seemed as though it wasn't real, as if it was one of the toy villages he had seen in stores.

Since it was also the first time Nick's father had flown

with his friend, the pilot apparently decided that he should show off by displaying some of his flying skills. He banked the plane to a hard left and then nose-dived about a hundred feet before leveling off. Nick screamed at the sudden and unexpected change of course and was suddenly so disoriented that he thought for sure he was going to die. He could hear his father and his friend laughing but he still didn't feel safe. Another quick jerk of the pilot's wrist and they were suddenly shooting directly upwards and all Nick could see was blue sky, with no earth anywhere in sight. He forced himself to turn his head to the pilot and without a second thought he reached over to grab the throttle in an attempt to re-right the plane. This time it was the pilot who panicked and yelled at Nick's father to take control of his son. The plane bobbed from side to side for several seconds before the pilot regained control and his father held Nick's arms by his side until they had landed.

Since that incident, Nick had not flown again and didn't even like being in a car while someone else was driving. For that reason, among others, he had been partnered with Inspector Tomas Castagna, who loathed driving. They were a perfect match.

Nick entered the precinct at ten fifteen and was greeted with a few smart comments and people glancing at their watches.

"Good afternoon, Nick," he caught from his partner while he was pouring himself a cup of coffee, "I thought you might have quit the force."

"Yeah, all right, that's enough Tomas, I work enough overtime. I'm entitled to a day off once in a while."

Nick was handed his mail and messages from the secretary assigned to homicide, and then he sat down behind his desk. He leafed through the pile of files and mail and noticed some reports that he had been waiting for had arrived from forensics.

He took a deep breath and laid them back on his desk. He wasn't ready yet for technical reading so he began scanning the Police Newsletter. Two new vehicles were being purchased for his precinct along with a copy machine and scanner.

Great, welcome to the 21st century.

He took another sip of his coffee and from the corner of his eye he saw his boss, Inspector Esposito, standing in the doorway.

Snapped out of his thoughts Nick said, "Damn it, boss, how long you been standing there?"

Esposito entered the office and took a seat across from Nick. "Long enough to see that you are only half here this morning. Rough weekend?"

Inspector Esposito was a bear of a man. Almost six and a half feet tall, pushing two hundred and twenty five pounds and at fifty years of age, he was probably in as good shape as many men half his age. Though he never liked to use his size to his advantage, he preferred listening and absorbing as much information as possible before speaking his mind, because he knew that when he spoke, people paid attention.

He once told Nick that his father used to say, "Son, your brain is stronger than your muscles and lasts longer too." It wasn't exactly Walt Whitman, but it worked for him.

Esposito had been with the force since he was in school,

his father had been an officer and got him a part-time job running errands on Saturdays. After high school he immediately entered the training academy and was soon a respected officer, not only by his colleagues, but by the people in his district. They called him *urso*, 'bear', for obvious reasons. Esposito wasn't particularly fond of the nick-name but he liked the people and, for their sake, let the name stick. He was both intelligent and diligent which was why, at the age of only forty, he was promoted to the Head of Homicide in the city.

"Did I have a rough weekend?" Nick said. "C'mon boss, I'm a family man. Sonja and I took the boys fishing. We had a nice time despite the weather. Alvar even caught a twelve-pound bass. It tasted delicious on the grill." Nick balled up some old notes and tossed them into the trash.

"Good, then you're all rested and ready for work?" He took a seat.

Nick looked at his boss who suddenly had the expression of a prosecutor preparing to plead his case. He clearly had something specific in mind that he wanted to talk about.

"What's up?" Nick said.

"I've got an interesting case for you."

"Oh, shit."

"What? I haven't even said anything yet."

"Yes, but the last time you told me that you had 'an interesting case' for me, Castagna and I ended up with a team of divers in the Tiete looking for a missing head. We ended up finding it, along with two other corpses we weren't even looking for. I can't stomach that shit. What ever happened to the clean and easy, single bullet murders? Give me one of those, boss."

"But they're no fun, no real detective work, it's usually the spouse or the ex. Besides, like I said, this one looks a little interesting and I want it handled right. You're the best guy that I've got." Esposito said with a cajoling smile.

"Sure, boss, I'm only up to my eyebrows in work, whereas everyone else is under water, is that what you're trying to say?"

Nick knew he couldn't turn his boss down but just wanted to toy with him a bit before the discussion turned serious, which it looked like it was about to.

Esposito cracked his knuckles and looked over Nick's shoulder out the window as he spoke. "Did you hear about the professor at the university who died last week, the mathematician?"

"No, I don't watch the news, too gruesome."
"Knock it off, Nick, I'm serious."

"Yes, I heard about it. Heart attack or something, died in his sleep. What's that got to do with us?"

"I don't know yet, maybe nothing, but the guy's life insurance company gave me a call and asked us to look into it."

"What am I missing here? It was a heart attack, right? And if I remember correctly, I don't think the guy was a spring chicken either. What's the angle?" Nick logged into his computer, only half interested in the conversation, thinking it was a no-brainer which just needed a few phone calls to close the loop.

Esposito pushed on. "The insurance company tells me that all the guy's paperwork was in order, you know, his will up-to-date, bills paid off, all investments in both his and his wife's names."

"Man, the guy died on Friday and by Tuesday morning the insurance company already has all this background work done? Don't those guys take a day off?" He scanned through his email. "So far, all it sounds like to me is that the guy was organized. Damn it, he was a math wizard, what do you expect?"

"Okay Nick, here's the kicker, the guy closed his bank account the morning of his death, transferring all holdings to his wife's and kids' accounts. He even finalized the paperwork detailing trust funds for his grandchildren, including a stipulation for grandchildren born *after his death*."

That finally caught Nick's attention.

"Who knows," Esposito continued, "maybe the professor died of natural causes, or....maybe he was blackmailed, or poisoned, maybe he had a heart problem and knew his time was limited, maybe lots of things, but something isn't right and we ought to look into it. More specifically, you ought to look into it."

With that, he tossed a file folder onto Nick's desk and stood up. "I've done some preliminaries in gathering family phone numbers, names of colleagues at the university, and tracked his bank transactions. An autopsy was performed yesterday; the results should be ready by this afternoon. Give me an update Friday. I want this one wrapped up quickly. We've got plenty of other work to do."

"No kidding boss, you, me and the insurance agents alike. I think we all need a vacation." Nick watched his boss walk out of the office, shaking his head at him as he shut the door.

Well, at least this case was different, not like his normal repertoire of 'murder and mayhem'. He sighed deeply and looked back to the stack of mail on his desk. He knew that he should take care of the forensic reports first but just couldn't bring himself to do it. They were going to have to wait until the afternoon.

Nick opened the file on the mathematician and was first confronted with a photo of the professor: clean-cut, medium-dark skin, full head of black hair with some graying over the ears. He looked healthy enough, but maybe it was an old photo. He turned it over and noticed a date scribbled on the back. Two months ago. Well, that was good, a recent photo can sometimes come in handy.

The next sheet in the file was a list of names and phone numbers including the insurance company and agent, the wife, kids, colleagues, and a few other names that may have been friends or neighbors. Nick picked up the phone to dial the wife and as he finished scanning the sheet he noticed the home address, it was in Sao Vicente, south of the city and towards the coast. Nick put down the phone, grabbed his keys and headed out the door.

Castagna caught a glimpse of him as he was hurrying past his office and called out. "Hey, where are you headed? Had enough of us already?" Nick ignored the comment and quickened his pace.

The sky was still overcast after the weekend rains and there were few beach-goers to slow down his travel. Nick was out of the city in thirty-five minutes. It felt good to head towards the beach. He felt like he was playing hooky. He probably should have let his partner know what he was up to, but for some reason he wanted a little time alone. Plus, he didn't think Tomas would think his plan was such a great idea.

Nick parked his car across the street from 56 Pasquala Place in Sao Vincente and noticed that it must be one of the few neighborhoods left in the area that had not been taken over by apartment houses catering to the tourists and beach crowd. The driveway was full of cars and, as he sat contemplating his next move, two more pulled in behind him. It looked as though the funeral had been that morning and friends and relatives were gathering at the home. For the family, right now was probably not a convenient time to have a police interrogation, but for Nick, the timing couldn't have been better. The house was full of potential interviews, and even suspects for that matter.

He got out of his car and started up the front walkway behind an older couple who had just arrived. The front door was open and with no one there to greet them, they began to mingle with the crowd. Inside were about two dozen people and Nick immediately began trying to determine who was who. There were several older women, any of whom could have been Senhora Rodriguez, along with a small herd of kids running about the house, one of whom had just bumped into him from behind.

"Sorry." The little boy said and was about to scuttle away when Nick grabbed him by the arm. "Hey, I said I was sorry. Let me go!" the little boy said as he squirmed under Nick's hold.

"Wait a sec," Nick said and bent down to the kid's level, trying his best not to cause a scene, "I just wanted to ask you a question. Can you tell me where Senhora Rodriguez is?"

"You mean mama? Aunt Maria? Or grandma?"

"Yes, I mean your grandma. Do you know where she is?"

Before the boy had a chance to answer, Nick noticed a new pair of legs next to him and looked up to be greeted by a smile from a middle-aged woman holding a tray of sandwiches.

"There she is!" said the little boy, smiling up at his grandmother, and immediately darting off when Nick's grip loosened.

"Hello," said the woman, "I'm not sure that we've met before, were you one of my husband's associates at the university? Or perhaps a student?"

"No m'am, I'm afraid not. I'm with the Sao Paulo police department."

"Oh, I'm sorry," she said with surprise, "were we making too much noise here, or is there a problem with the cars on the street?" She turned her head to look out the front window to evaluate the potential traffic problem.

"There's no trouble at all with the gathering. Actually, I'm with the homicide department." He then reevaluated his timing, arriving at the house just after the funeral. "It looks like I've come at a bad time. I wasn't aware that you had a house full of guests."

"One more won't hurt." She replied. "Did you know my husband?"

She must have misunderstood his presence. "Actually, I didn't. I'm here because I've been asked to look into your husband's death."

The woman was clearly taken aback and handed her tray to another woman walking by. "I'm sorry Inspector..."

"da Silva."

"Inspector da Silva. There must be some misunderstanding. My husband died of a heart attack in his sleep." She gently took hold of Nick's elbow and guided him through the living room and into what appeared to be a small home library, she then shut the door behind them.

"I found him myself when I tried to wake him in the morning. He was laying there with such a peaceful look on his face. I thought he was having a pleasant dream, but then I touched his cheek and it was cold. I knew right away that he was gone."

"Senhora Rodriguez, I'm sure there is nothing to be alarmed about, but your husband's life insurance company has asked us to look into his death."

"I don't understand. Is there a problem?" she asked. "Oh... I haven't even thought about that yet. There is so much to do!"

"Actually, at this point there is probably nothing to worry about. The insurance company has reported that his estate is entirely in order and up-to-date."

She smiled and then, as though she were speaking to her late husband, she said, "Oh, Eduardo, everything must always be in order."

"My husband was a mathematician," Senhora Rodriguez went on. "Everything had to 'add up' so to speak, so I'm not surprised that his paperwork was so well organized."

"Well, it was apparently more than just organized," Nick said, "we usually find this type of order only with terminally ill patients who are preparing for death."

"I... but my husband wasn't—"

"May I ask, was your husband ill?"

"No... well, nothing to speak of really, surely nothing that should have killed him. He did have slightly high blood pressure, his doctor told him about six months ago to watch his diet and get some exercise, but said that medication was not necessary at that point. He was supposed to see the doctor again this week. Oh, if only that appointment had been last week, Eduardo might still be with me." She was clearly shaken and began wringing her hands in her apron.

Their attention was then drawn to the door as it was quietly cracked open. A younger woman, perhaps in her mid-twenties, with thick black hair and large dark eyes peered into the office.

"Mama, is everything all right?" She had the same soft voice as her mother.

"Yes, Maria, everything is fine. Come in. This is Inspector da Silva."

The daughter slipped into the room and shut the door behind her.

"Apparently papa was so well organized that the insurance company thinks he knew that he was going to die. Isn't that so like your papa?" Senhora Rodriguez said to her daughter.

Nick decided that he should end the discussion. She had enough to digest at the moment.

"Senhora Rodriguez, I don't think this is the best time to discuss this issue, would it be possible for me to come back tomorrow and talk further, perhaps with your children also?"

She took a deep breath. "I'm not sure that there is anything more to discuss Inspector. But I suppose we should straighten out any misunderstandings with the insurance company as soon as possible, especially after Eduardo took so much care in preparing everything for us. Why don't you come by tomorrow morning around ten o'clock and I'll see that Maria and my son Miguel are both here also." She looked to Maria who nodded in agreement. "Maria, would you please show the Inspector to the door?"

"Sure, mama." Maria opened the office door and motioned for Nick to follow her.

"My condolences on your loss, Senhora Rodriguez." Nick said to the professor's widow in parting after realizing that he'd forgotten the formality earlier.

"Thank you," she replied with a graceful nod as she watched the Inspector leave the room with her daughter.

As Nick followed Maria back through the living room amongst a considerable throng of grievers, he couldn't help but notice the simplicity of the home. There was little decoration other than the built-in bookshelves on the long interior wall, which was adorned with small photos in frames interspersed among the volumes of books. A tapestry hung above the couch displaying a desert scene with palm trees, camels and some Nomadic tents. A single oriental rug lay on the floor in front of the couch upon which a plain, dark-wood coffee table stood. Two arm chairs were positioned on the opposite side of the room between which was a small table with a reading lamp and a few magazines stacked purposefully on one edge. The room was certainly orderly, but despite its near barrenness, it exuded a welcoming, homey warmth.

At the front door Maria stopped and turned to Nick.

"We will see you in the morning, Inspector." She nodded with a respectful smile then turned back to the guests in the living room.

Nick walked out the door behind a man in a beige tweed suit and small bow tie. He paused for a moment on the front steps, and noticed the man in tweed now standing with another gentleman smoking a cigar.

"Did I hear Maria say that you are an Inspector?" asked the man in tweed.

"Yes, my name is da Silva and I'm with Sao Paulo Homicide."

Nick thought perhaps he could get some additional information from the two men before he departed so he confirmed his homicide standings to get their attention.

It worked.

"Homicide? Wow, tough job," said the man who was being polite without showing much enthusiasm.

"I don't suppose you are here on business? Ha, ha!" said the man smoking the cigar who looked to his companion for a supportive smile.

"Actually, I am. I'm investigating the Professor's death."

Both quickly turned to Nick.

"Anything the two of you can add could be helpful."

Nick watched as their brains were apparently working a mile-a-minute to comprehend what he had just said.

"But... but I thought that he died in his sleep," said the man in tweed.

Nick loved these games. "That doesn't mean that he died of natural causes, does it?"

Put on the spot, the man scrambled for a quick answer. "Well, no. No, not necessarily. I mean... I'm sorry. I don't understand. Why is his death being investigated?"

"May I ask your names and relation to the deceased?" Nick asked while taking out a small notepad to jot down the information.

"I'm Professor Montagna," stated the man in tweed, then gestured to his companion, "and this is Professor Kumar. We both work, or rather *worked*, with Professor Rodriguez at the university."

Nick continued the impromptu interrogation. "Have either of you noticed a change in Professor Rodriguez lately?"

Both were clearly trying to recall if they'd noticed anything over the past several months.

Finally Professor Kumar spoke. "What kind of a change are you suggesting?"

"I'm not *suggesting* anything at all. I'm merely trying to determine the facts." Both men were becoming noticeably agitated.

Nick was really beginning to enjoy himself and realized that this may not be such a bad day after all. "Did he act differently in any way? Distracted, scared, or perhaps feeling a bit under the—?"

"Wow!" Professor Kumar interrupted. "Do you think that someone killed him? I mean that's what homicide investigations are all about." Kumar finished his thought and then turned to his companion to see if he had any thoughts to add.

It was Montagna's turn next and he spoke as though he were choosing his words very carefully. "Inspector, I think that both of us would be happy to help you in any way that we can, but Professor Rodriguez was a very private man, neither of us spoke with him about much more than our teaching, and even those occasions were few and far between. I can't say that I noticed any kind of change in him recently, but if you could give us some more ideas as to what we should be focusing on, then possibly that would jog our memories."

Cutting to the chase Nick asked, "Did Professor Rodriguez appear to be ill?"

"Not at all," replied Kumar, "in fact, he was always so calm and seemingly without stress that I was surprised to hear that he had suffered a heart attack. Perhaps there was more to the Professor than we knew."

Nick decided to end the conversation with a little more fun. "Perhaps so. Tell me, do either of you know anyone that may have wanted him dead? A vengeful student? Scorned lover? Or maybe even a jealous colleague?" He waited for an answer while trying not to smile at their expense.

Kumar was the first to respond. "Professor Rodriguez was a model of a human being, a brilliant researcher, an excellent teacher, great with the students and was, as far as I knew, devoted to his family. I can't imagine anyone wanting to harm him."

"I completely agree. I can't think of a possible circumstance where someone would want to take his life," Montagne said.

"But how can you both be so sure? I thought you just said that he was a very private man."

"True, however—" Kumar began.

But Nick decided to end the game abruptly. "Well, I'm in a rush. Thank you both for your time," Nick said shaking both of their hands, "and I'll be in touch."

Nick turned and walked down the drive, leaving the two to create their own theories about who may have killed the esteemed professor. He knew he'd be meeting them both again and would be looking forward to seeing if either would have anything to add to their stories.

Chapter Five

Ron and Caroline drove together in silence. It wasn't as though they weren't open to conversation, but lately every time they broached the subject of her illness they both broke down in tears, so sometimes it was just easier to remain silent. Ron pulled the car into the parking lot outside of Dr. Mansfield's office and turned off the ignition.

"Are we ready?"

Less than optimistic, "I'm ready," Caroline replied, "are you?"

This was the first time that Ron would be speaking to the doctor about Caroline's leukemia. It would be difficult for everyone.

"Sure. Let's go."

Dr. Mansfield's office was located just outside of Wakefield, less than 20 minutes from their home, and was one of several physicians' offices found in a red-brick building originally constructed as an out-patient clinic for South County Hospital. The sidewalks leading around the building and into the front entranceway were lined with small boxwoods, surrounded by crocuses which were just starting to bloom, little buds of white and varying shades of purple.

Ron held Caroline's hand tightly as they walked into the

building and took the stairs to the second floor. He had never formally met Dr. Elaine Mansfield, but knew of her not only through Caroline and several other friends for whom she was their physician, but also because they had attended the same high school. Ron was three years younger than she and although he didn't remember her from their school days, Caroline had found a few photos of Dr. Mansfield, as a senior, in Ron's freshman high school yearbook. She was apparently quite active: a member of the field hockey team, Amnesty International, the Biology Club, and editor of the school newspaper just to name a few. In all the photos she had her long hair pulled tightly back into a pony tail and wore glasses with odd square-shaped rims. She looked as though she loved having her nose stuffed in a book. The perfect candidate for medical school.

Caroline told the receptionist that they were there and then they sat down in the waiting room on a small leather love seat, which was the only free space available. Ron counted, there were eight others waiting.

Just as they had resigned themselves for a long wait, the receptionist called them back into the offices. They were led down a wide hallway which was decorated with several pastel-shaded watercolors brightening the walls and hopefully the moods of the patients.

Ron was surprised to find that in Dr. Mansfield's office there was no examination table or equipment, but rather a long wooden desk in a light oak with contemporary curves and matching chairs. Behind the desk were the standard bookshelves filled with medical reference books, file folders and a few photos. There was a crystal paperweight on the desk in the shape of a penguin next to which lay a memo pad on which was printed "Don't Touch. Doctor's Orders".

They heard the office door close behind them and Ron turned around in his chair and began to stand up.

"Oh, please stay seated. You must be Ron. I'm Elaine Mansfield."

"It's a pleasure to meet you, doctor." Ron shook her hand as he retook his seat.

As the two women were exchanging greetings, Ron couldn't help but notice that Dr. Mansfield no longer looked like she did in her high school photos. Her blond hair was still long but it was now pulled back into a twist with a few strands falling loosely about her face, giving her a very warm, approachable appearance. The odd-shaped glasses were gone, undoubtedly replaced by contact lenses. He realized that she was quite striking even though she was wearing very little make-up, simple gold stud earrings and no other jewelry.

Just then Caroline gave his hand a squeeze and looked at him with a wry smile, she must have read his mind. He smiled back at her and realized that he felt comfortable there, as though they were in good hands. Off to a good start.

"So, how are you feeling, Caroline?" Dr. Mansfield began, apparently skipping the small talk.

Caroline began to get out some of her bottled-up emotions. "Well, physically I feel pretty much the same, some days I feel perfectly normal, and others I'm quite exhausted."

"And emotionally?"

"Emotionally, I feel confused. I have so many questions from 'Why me?' to 'What now?'"

"Well, that's why we're here. We'll take all the time you

need, so just relax. First, why don't you let me explain your condition to you again so that Ron... is it alright if I call you Ron?"

He nodded.

"Good. Yes... so that Ron can hear everything which we have already discussed and then we can start tackling your questions. Okay?"

"Okay." Both Ron and Caroline said in unison and Dr. Mansfield began.

"Ron, when your wife came to me complaining of fatigue and a cold that she couldn't get rid of, I gave her a blood test. The results of the test indicate that she does indeed have leukemia. A further bone marrow test revealed that she has likely had the cancer for many years and it is just now beginning to display symptoms."

Dr. Mansfield turned to Caroline, "The cancer which you have is called *chronic lymphocytic leukemia*, this type is most commonly found in adults and over eight thousand people will probably be diagnosed with it in America in this year alone. Most cases are found in men over fifty, but it also appears in women and in your age group too, though rarely."

"That's the 'why me?' question." Caroline said in almost a whisper.

Ron put his arm gently around her shoulders.

"I'm sorry, Doctor Mansfield." Caroline said, "Please continue."

"Alright. Now, I'm not sure how familiar you both are with leukemia in general so I'll start with the basics."

"Please do." Ron wanted her to start right from the beginning. Up to this point in his life he was unaffected by cancer. No one in his immediate family had been subjected to it. Cancer had always been something that happened to other people. He felt his stomach tighten and took a deep breath to try to relax.

Dr. Mansfield continued, "Leukemia is cancer that originates in the bone marrow, which is the soft, spongy inner portion of bones, and in which white blood cells, the leukocytes, are the malignant cancerous cells." As she spoke, she used a bone model to show all the inner structures. "Leukemia develops when a leukocyte undergoes a transformation into a malignant cell, a cancerous cell, then this malignant cell is capable of uncontrolled growth. Leukemia cells begin to multiply in the marrow, and as they do so they crowd out the normal blood cells. Those normal blood cells are the ones which carry oxygen to the body's tissues, fight infections, and help wounds heal by clotting the blood."

"This is why Caroline has not been able to get rid of her cold, because she doesn't have enough healthy blood cells to fight the infection. Leukemia can also spread from the marrow to other parts of the body, including the lymph nodes, brain, liver, and spleen—"

"Has that happened to Caroline?"

"No, Ron, not yet. At least there don't appear to be any signs of spread."

Thank God.

"Leukemia is found in two forms—acute and chronic. Among children, one form, called acute lymphocytic leukemia, accounts for about two-thirds of cases, while acute myeloid leukemia and chronic lymphocytic leukemia, which is what you have Caroline, are the most common types in adults." She paused.

"Am I moving too quickly for you?"

"No, please continue," Caroline answered.

"In acute leukemia, the malignant cells are immature cells that are incapable of performing their immune system functions. The onset of acute leukemia is rapid, and, in most cases, fatal unless the disease is treated quickly—"

"But that's not what I have is it?" Caroline asked.

"No, you have a chronic type of leukemia, which develops in more mature cells. These cells can perform some of their duties but not well and they increase at a slower rate, so the disease develops more slowly than in acute leukemia."

"So we have more time to find a cure for Caroline since she has the chronic type?" Ron thought he saw a window of hope.

"Well, yes and no. She does have the chronic form, but her leukemia has been progressing over many years and appears to be on the verge of entering the acute stage."

Ron's hopes plummeted.

"But, acute leukemia tends to respond better to chemotherapy, and with such a rapid growth rate, a bone marrow transplant is almost always attempted as soon as possible; whereas with the chronic form, chemotherapy is utilized first and often sends the cancer into remission allowing the patient to live without symptoms for an amount of time. For Caroline I recommend we begin chemotherapy as soon as possible while simultaneously planning for a bone marrow transplant."

Caroline and Ron looked at each other and Ron could see tears welling in his wife's eyes.

Chemotherapy was going to be rough. Everyone had heard about potential side effects of the treatment. But a transplant, too?

Ron asked, "Well, uh, what's involved in a transplant?"

Dr. Mansfield smiled cordially and continued. "In a bone marrow transplant, stem cells—the immature blood cells that give rise to all types of blood cells about which we've heard so much in the news recently—are harvested from the bone marrow or circulating blood. Most patients then undergo a preparative regimen, which is a course of high-dose chemotherapy or radiation therapy or a combination of both. The preparative regimen suppresses or destroys the cells in the bone marrow at the same time that it targets cancer cells. Shortly after the completion of this phase of treatment, the harvested stem cells are then infused into the bloodstream. Over the following days the stem cells migrate to the marrow space, and within weeks and months begin production of the full range of blood cells needed for normal functioning."

"That sounds simple enough, but why do I have the feeling it's not?" Ron asked.

"You're right Ron, it's not simple."

"For the donor or the patient?" Caroline asked.

"I was referring to the donor, although there are certain types of cancers in which stem cells can be harvested from the patient's own bone marrow or circulating blood and re-infused into the bloodstream; this is what is called an autologous bone marrow transplant. Unfortunately this type of transplant has not proven to be effective in treating your form of leukemia, you would require a donor to supply stem cells in what is referred to as an allogeneic transplant."

"Can anyone donate these cells?" Ron asked. Of course he would be more than willing to make the donation himself.

Dr. Mansfield sighed. "Unfortunately not, and that's where one of the biggest obstacles occurs, a perfect, or even partial match, is difficult to find."

"What does 'match' mean? What exactly are we matching?" Caroline asked.

"We're talking about genetic matching. Allogeneic transplants are least likely to have complications such as graft-versus-host disease and graft rejection when they come from a related, genetically matched donor."

"Like a sibling?" Caroline said.

"Yes, the ideal donor is a sibling who has inherited a specific set of genes identical to those of the patient. These genes express proteins called human leukocyte antigens or sometimes referred to as HLA molecules. These HLA molecules are found on the surfaces of white blood cells and help the body distinguish between its own cells and foreign invaders that should be destroyed."

Dr. Mansfield leaned forward on her desk and folded her hands together. "The chance that a patient and his or her sibling will inherit a matched set of HLA genes is just one in four. Only about a third of patients who could benefit from a transplant, however, have this ideal donor and meet other eligibility criteria. For the remaining 60 to 70 percent of patients, physicians expand the search to other family members who may be only partially HLA matched."

"And if no match is found in the family?" Ron asked.

"If there is no matched family member, the search extends to donor registries like the National Marrow Donor Program, which maintains a database of potential marrow donors and is linked with other national and international registries, creating a combined pool of 4.5 million potential donors."

"Four and a half million? Then we will certainly find a match with such a large donor pool, won't we?" Caroline asked.

"Hopefully we will, but first we will test your family and see if we can find a match there. In addition to your parents, you have a sister and a daughter, correct?"

"Oh my gosh, Sophie," Caroline turned to Ron, "I would rather not bring her into this, she's only four years old."

"Statistically speaking, you have a higher chance of having a perfect match from your sister, but biological children always provide a partial match because they receive half of their genes from each parent. If by some rare chance the parents have a common gene then the chances increase that an acceptable match exists."

Ron looked at his wife, "I would like to shield Sophie from this too, but you know she would want to help you if she could." Then he turned to Dr. Mansfield, "What is involved with the testing and eventual donation? Is it a complicated surgery?"

"The test for a genetic match is quite simple, it requires only a small blood sample to determine tissue type. The marrow collection process is naturally more involved. It is a surgical procedure lasting approximately one to two hours. The procedure occurs in a hospital operating room while the donor receives local or general anesthesia. Part of the marrow is removed from the back of the pelvic bone using sterile needles and syringes. Most donors recover quickly from the procedure and may have some bone pain and aches for several days or a few weeks but the marrow naturally replenishes itself within four to six weeks."

Dr. Mansfield looked at Caroline. "I understand your concerns about Sophie, we'll try to find another suitable match, but if she is the only match available and goes through the procedure, she will be just fine. Most children are excited about helping, especially if it's one of their parents, and they find the hospital experience an adventure, plus they usually recover more quickly than adults. Don't worry."

Ron took a deep breath and tried to collect himself. There was so much information that he felt completely overwhelmed and helpless. This was his family and there was little that he could do to control the situation. Not to mention that having his wife's life at stake was a nightmare.

Dr. Mansfield went on to talk about the chemotherapy sessions, combination therapies, types of drugs used, potential side effects and the importance of maintaining a healthy diet. Staying active and seeking medical attention for any ailment, no matter if it's just a common cold, and to be sure to have it treated as quickly as possible was very important.

"Now that you have some idea of what you'll have to go through over the coming months, do you have any more questions? Or would you like to meet again in a couple of days, once you've had a little time to think all this through?"

Ron quickly interjected, "I don't think we should waste any time if the cancer is progressing as rapidly as you say."

"I agree," answered Dr. Mansfield. "First we'll have to

schedule a visit for you with an oncologist for consultation and the preparative treatments."

"I do have one question," Caroline began, "if I don't find a suitable donor, are there any other options for a cure besides a bone marrow transplant?"

"To my knowledge, I'm afraid there aren't any others that have proven more effective. Of course there are several treatments in the testing stages. But chemotherapy alone has not been shown to be a cure, although it does slow the growth rate of the cancer cells. There are some procedures that we can implement to help your body in accepting bone marrow from a partially matched donor, but your case is in a relatively advanced stage. A bone marrow transplant is really our only option at this point. But as you know, I am not, technically, an expert in this field, so you'll have to discuss this with the oncologist, and I can refer you to a couple of other specialists who have more extensive experience with the treatment of leukemia."

"But Doctor Mansfield, you seem to be an expert," Ron said.

Dr. Mansfield hesitated a moment before continuing. "Ron, I told Caroline previously that my brother-in-law died of leukemia last year, which is why I am up to date in the diagnosis and treatment of it. He underwent a transplant with a partial match but the cells were rejected and he passed away before a new donor could be found. This disease can be beat, but it's not going to be easy. I'll do everything I can to help get Caroline through this."

"I appreciate your candor Doctor Mansfield," Ron replied.

"Caroline feels completely comfortable in your hands, and after our brief discussion today, so do I."

"I'm glad to hear that."

"Of course, I would like to consult some experts in this field," Ron said, "but I think Caroline and I would both agree that we would like to continue to work with you as her primary physician if you are willing to oversee her case." He looked for concurrence, "Do you agree?"

"Yes," Caroline said. "I think this is going to be difficult enough as it is. I feel as though I'm in good hands with Doctor Mansfield and believe that if anyone can help me, she can. At least she can point me in all the right directions."

"Well, I'll do all that I can," Dr. Mansfield said. "So, if you don't have any more questions for me today, I'll get to work on finding a donor for you. Then we'll try to get you back into good health as soon as we can." Dr. Mansfield stood and walked around the desk to shake Ron's hand, she then turned to Caroline and gave her a warm hug. "I have a good feeling about you, Caroline."

Ron looked at his wife and saw tears forming in her eyes again. He led her to the door.

"Please make an appointment at the front desk for the blood sampling for your family," Dr. Mansfield said. "And stay positive, we'll see each other again in a few days."

"Thank you, Doctor," Ron said as he escorted his wife from the office.

Elaine Mansfield shut the door behind them and returned to her desk. She sat down, opened one of the desk drawers and pulled out a photo of her sister with her two boys and husband who had died last year. What a cruel disease this was.

The lives of her sister and nephews will never be the same without him.

Years of studying medicine and she was completely helpless in saving his life. The feeling of loss was still hanging heavy in her heart. She carefully placed the photo on her desk so that she could see the family during their happy days.

She stood up and looked out the window overlooking the parking area where she saw Ron and Caroline walking to their car. Ron guided his wife to the passenger side, then took her in his arms and held her for a moment, giving her a gentle kiss on the forehead before opening the door for her.

Dr. Mansfield turned away from the scene, wiping a tear from her cheek as she gathered her files together to meet with her next patient.

Chapter Six

Rosalinda Rodriguez sat with her two children and daughter-in-law at the breakfast table. The three grandchildren were on the veranda playing with a rocking horse, which had belonged to their father when he was a child.

"Would you like something more to eat Miguel?" Rosa asked her son.

Miguel pushed his empty breakfast plate away and picked up the file folder that was laying in front of him. "No thanks, mama. I think we need to talk before the police inspector arrives."

Rosalinda knew that she could not put off the conversation any longer. Miguel had wanted to discuss the estate settlement last night but she had insisted that she be left in peace to grieve for her husband on the day of his funeral. She didn't have much interest in financial matters and was hoping that Miguel could handle everything himself. He was the head of the family now, and she believed that it was his place to not only look after his wife and children but his mother and unwed sister as well. "I don't see why I need to be involved in this Miguel. Couldn't you handle the paperwork by yourself and just let me know if you need my signature somewhere?

You know better how to deal with these matters, and I would just get in the way."

"Mama, it's not that simple. I've already discussed this with Maria and she agrees that we need to talk with you."

Rosalinda turned to her daughter for confirmation and could see from the look on her face that she was in agreement with Miguel.

"Yes, mama," Maria said, "Miguel explained some things to me and I can't make sense of it all. It's as if what the Inspector said to us yesterday is true, that papa knew that he was going to die, perhaps not so soon, but he at least knew that he had very little time."

One of the boys on the porch began to cry and Miguel looked to his wife Katie, who excused herself to tend to the children.

Rosalinda could not even begin to comprehend why they believed this to be true. The idea that her husband could have known he was going to die and did not talk to her about it was unthinkable, but before she could begin asking questions, Miguel started to explain about the file folder that lay before him, which he had found in the top drawer of his father's desk. It contained a copy of his father's last will and testament, which was dated on the day before his death. In addition to the will, there was a complete itemized list under the names of Rosalinda, Miguel and Maria detailing personal possessions, which had been respectively bequeathed to each of them.

Miguel began, "At first I thought that papa had simply been very thorough and up-to-date and that it was just a coincidence that he had worked through these papers on the day before he died."

Rosalinda could see that Miguel was struggling with his thoughts. "Go on," she said.

"But then," Miguel continued, "I called my bank to ask about handling the trust funds that papa had left for my kids and they notified me of a third party deposit that had been made into my account on Thursday. Apparently father had deposited a large sum of money into my account, which happens to coincide exactly with the sum in his will under my name."

Maria seconded Miguel's claims, "I did the same once Miguel told me what he found out. Papa made an equal deposit into my account and he closed his own account on the same day!" She began to cry and between her tears she murmured, "If Papa knew that he was dying, why didn't he tell us? Why didn't he say goodbye?"

Then her eyes widened, "Or what if... what if he killed himself?"

Rosalinda reached across the table for her daughter's hand and held it saying, "Oh, darling, you know how much papa loved us, he would never have done anything to hurt us. Please don't get upset. He could not have known he was dying, nor did he have reason to end his life. He would have talked to me about it, I'm sure of it. What Miguel has said could easily be coincidence."

"There's more Mother," Miguel said. "The funeral home that we used for Father's burial wasn't selected by me. I was planning to call several locations in the area to compare prices when the phone rang on Friday afternoon. We had just arrived and you were outside with the kids so I took the call. It was the Sao Francisco Funeral Home and they said that they had found out about father's death through the coroner's office. They went on to explain how they already had an account set up in papa's name to pay for all of his funeral and burial arrangements. They told me that it was not standard practice for them to accept such arrangements except with the terminally ill, as most people avoid thinking about such formalities until required. This occurred two weeks ago and the director said that everything was arranged through a courier who was sent to the funeral home with the appropriate sum in cash and a description of services desired."

Until that point Rosalinda was successful in maintaining a calm exterior, but now she was deeply troubled and her hands began to shake. "Why... why didn't you mention this to me then?" she asked.

"At the time, I didn't think much of it. Father was a planner, so I thought he was taking care of this as a precaution for the future. He has always gone out of his way to care for all of us, especially for you and Maria since I've moved to Brasilia. It wasn't until I found this file and talked to the bank that I put two and two together." Miguel held the file folder up as he talked, as if it were solid evidence for his case.

"It doesn't make any sense." Rosalinda shook her head. But then she slowly began to understand. "So this must be what the Inspector wants to talk about. The insurance company must think that because of his careful planning, he knew he was going to die."

Miguel took a deep breath then added, "Or at least *someone* knew that he was going to die."

"What is that supposed to mean?" Maria asked, suddenly jumping back into the conversation. "*Someone*? Like who?"

"Like someone who may have been blackmailing him. I don't know. I don't really know. I'm as confused as you two and I'm just trying to figure this all out."

"Well, maybe the Inspector can help us with that," Maria concluded. Turning to Miguel, she added "Do you think we should show him father's paperwork?"

"If he's worth his two cents in investigative work, then I'm sure he can easily get a copy of father's will and probably already knows about the funeral arrangements. I think we should be honest with him. We have nothing to hide and I for one would really like to know what is going on here."

Rosalinda nodded. "Let's give the Inspector everything that we have and all that we know and get this taken care of as soon as possible. Poor Eduardo would hate to see us going through this."

She stood and walked into the kitchen where she lifted the coffee pot off the stove. She was trying to process all that she had just heard, but it was too much. All she wanted was Eduardo to be there with her, to tell her not to worry, that everything would be alright. She returned to the table and filled her children's coffee cups. "I'm sure there is a simple explanation for everything."

But she didn't even believe her own words.

Katie returned to the kitchen and announced that Inspector da Silva and another officer had arrived and were waiting in the front hall.

"Why don't we bring them into the living room to talk,"

Miguel said to the others. "Katie, would you like to join us for this?"

"No, I'll watch the children so you can have some peace and quiet, but call me if you need anything," Katie responded.

Miguel walked up to his wife and gave her a kiss while patting her already noticeably pregnant stomach. She was only three months pregnant but since it was their fourth child, her body seemed to be used to the routine and began filling out almost from the start.

Rosalinda and Maria followed Miguel into the foyer where the two men were standing in quiet conversation. Miguel shook hands with the officers.

"You must be Inspector da Silva. I'm Miguel Rodriguez."

"Pleasure to meet you, Mr. Rodriguez," Nick responded. Then he nodded towards his partner and said, "This is Inspector Castagna. We'll be working on this case together."

"I see," Miguel replied. "And you've already met my mother and my sister?"

"Yes, I've already had the pleasure." Nick nodded to the two women. "Thank you for meeting with us again at such a difficult time," he said with sincerity.

"Please, come inside where we can sit down and be comfortable." Miguel motioned the group into the next room.

"Would you like coffee, Inspectors?" Maria asked.

"No, thank you. I'm fine," Nick responded, as did his partner.

"Well then let's get right into things and get this over with," Miguel said.

"Right," Nick began. "As I mentioned to your mother and sister yesterday, your father's life insurance company has asked the Sao Paulo Police Department to look into your father's death because they believe they have evidence that it was perhaps... well... anticipated."

"That is ridiculous," Rosalinda said.

"Mother, please." Miguel turned to her and gave her a firm look.

Rosalinda regained her composure, but was clearly not planning on giving much credence to the conversation, which she believed to be a waste of her time and emotional energy.

Nick continued. "As I'm sure you are already aware, your father had his last will and testament updated on the day before his death. He liquidated his holdings, transferred—"

"We're aware of all this, Inspector." Miguel was also on edge. "Could we please move on to how we can help you?"

"Of course," Nick continued. "Senhora Rodriguez, I know that I asked you this yesterday, but I must ask again. Was your husband ill? Is it possible that he was terminally ill?"

Rosalinda's hands were clenched tightly in her lap. "As I already told you, his doctor said he should watch his diet because of slightly high blood pressure, but that's all. The doctor assured both of us that it was nothing to worry about. Apparently, he was wrong."

Miguel had always known his mother to be such a sweet, warm, soft-spoken woman but now she wasn't acting like herself. Of course she was grieving over the loss of her husband, but she also seemed personally assaulted about the insurance issues and a possible health problem that his doctor had missed. She was apparently trying to hide her feelings from her family, but he knew her too well. It hurt him to see her like this and he wanted to shield her from any further pain.

"Inspector, as far as we know, my father was in excellent health." Miguel added to his mother's comments. "Perhaps his doctor could shed more light on that subject, but I'm afraid that we can't."

Nick changed the subject. "I believe you live in Brasilia, correct?"

"Yes, that's correct."

"When was the last time that you saw your father?"

"At Christmas. He came with my mother and sister to visit us in Brasilia."

"And you noticed no change in his demeanor at that time?"

"No. I did not."

"What do you do in Brasilia?"

"I'm an architect, but I have a feeling that you already know all this." He hesitated a moment and then added. "As I'm sure you know about the funeral home arrangements."

"Arrangements? Is there any reason that they should interest me?" Nick asked.

Miguel went on the describe the phone call that he received from the Sao Francisco Funeral Home while Inspector da Silva jotted down some notes.

"I see," Nick responded. "Is there anything else that you would like to tell me about? Anything out of the ordinary?"

"It's hard to say what's ordinary, Inspector. This is the first time that we've had to deal with an immediate family member's death." At Miguel's words, Maria began to cry softly.

Nick turned to her and asked, "Miss Rodriguez, did you notice your father's behavior at all different prior to his death?"

Maria looked up at him and touched a tissue to her moist eyes. "No. No. He was the same sweet, loving papa that I've always known." She put her face in her hands and continued to weep.

"What about you Senhora Rodriguez? Was your husband acting at all different recently, especially on the night that he died?"

"No, not at all. We had a wonderful evening together. As we always do. This came as a total surprise," she answered.

Miguel added. "I even spoke to him on the telephone on Thursday and he acted just as he always did, except maybe that he got a little emotional when I gave him the news that we are expecting another child." Then almost to himself, "I'm so glad I told him. I'm so glad he knew."

Then Nick asked Miguel, "Did your father call you or did you call him?"

"Actually, he called me at work. We normally speak once a week, usually on Sundays so that he can talk to the kids too, so I guess his call was a little unexpected, but nothing that made me think something was wrong."

Then Maria spoke up. "He called me at work on Thursday too, which he never does. I work at the newspaper and private phone calls are really frowned upon."

"Go on," Nick said. "What did he want to talk with you about?"

"Nothing in particular. He asked me how my job was going and then told me that if I got married, then I could stay home and not have to worry about earning money." She finally smiled. "I told him I'd think about it. We liked to tease each other."

"Is that all?" Nick asked. "Is there anything that you may have overlooked?"

"No. Well, yes, actually, he invited me to come over for breakfast on Friday morning, at eight-thirty." Her eyes filled again with tears. "Oh, how I wish I was able to have seen him one last time! But I arrived just as the coroner was taking him away."

Rosalinda placed an arm around her daughter's shoulders and proceeded with her story. "Since Miguel would have to fly in, I called him as soon as I realized Eduardo was gone. But I knew Maria was coming over and I wanted to tell her in person."

"What time did you call your son?" Castagna spoke up for the first time.

"About six. Miguel told me to call an ambulance, which I did. I told them on the phone that Eduardo was cold. I knew that an ambulance would do no good, he was already gone." Rosalinda began to cry now too, but continued. "Then I called Father Angelo who arrived just after Maria. Thank goodness he was here for us. He has been such a blessing to us the past few days."

"From which parish is Father Angelo?" Nick asked.

"Saint Francis de Sales, here in Sao Vincente. It's a wonderful congregation. We've been attending services there since we were married, over 35 years now." Rosalinda was clearly proud of her church and her faith, a respite in her sorrow.

Miguel smiled at his mother. He loved her so much and knew that he needed to bring this interrogation to an end. His mother needed the comfort of her family now. He turned to the Inspectors. "I'm not sure if there's any more information that we can provide you with. So, if there are no more questions I think we are through here."

"All right," Nick said. "Thank you for your time Senhora Rodriguez." Then turning to Maria, he added, "Miss Rodriguez, I'm sorry for your loss."

The two Inspectors followed Miguel to the front door, then Nick asked to speak with Miguel for a moment outside.

But on the front steps Miguel spoke first. "I'm really not sure where all of this is going. If my father was sick and knew he was going to die, none of us knew about it. I seriously doubt that was the case." He was turning things over in his mind. "I know that my mother has no financial problems now, but she is still relatively young and will not be able to live off her savings forever. Is there going to be a problem for her to collect my father's life insurance money?"

"Perhaps. I'm not sure yet." Nick turned his head towards the street then added. "Mr. Rodriguez do you think it's at all possible that someone may have wanted your father dead? Could he have been a target of blackmail?"

Miguel was now so confused and tired of trying to figure things out. "Inspector, I honestly don't know what to think. If you had known my father, you would know that it is an impossibility for such a man to have enemies."

"I appreciate your candor, but unfortunately, I didn't know your father, and apparently the insurance company didn't either. We are just doing our jobs."

"I know," Miguel said, "but I really think that you are way off base here, unless someone was envious of his work. But even then, I can't fathom that anyone could want to hurt him.

And on top of that, I don't want this to get difficult for my mother. I'd be happy to help you for the sole purpose of easing the burden on her, but beyond that I am not going to postulate any foul play because I believe it to be an absolutely impossible scenario."

Nick tried to understand the feelings of the man before him, but too many times he had looked into the eyes of a guilty man and had been tempted into believing his lies.

"I'll do everything I can to wrap this up as quickly as possible," Nick said, "but I can't promise that the foul play theory will disappear without further investigation."

"I guess I can't ask for more." Miguel reached into his wallet. "Here's my business card. I need to return to Brasilia tomorrow for a conference. Please notify me prior to contacting my mother, if possible."

"I will... if possible." Nick began heading down the front walkway with his partner at his heels. "Good day, Mr. Rodriguez."

Miguel gave no response, then turned and reentered the house.

Nick felt that dealing with family members was one of the most difficult aspects of his job. Sometimes they are so emotional that they are of very little help in providing useful information, as was the case today. He felt that the Rodriguez family, at least the women, were likely telling him the truth, but they may have overlooked something that could help, perhaps a slight change in the Professor's behavior, an off-hand comment implicating someone else's involvement, or possibly a subtle way of saying goodbye.

Nick felt in his gut that something was askew and when

he got that feeling, he was usually right. A piece of the puzzle was missing, he was sure of it, and he intended to find out who or what it was.

And then get some answers.

Nick and Tomas got into their car and Nick steered towards the center of Sao Vincente.

"Where to now?" Tomas asked.

"The priest."

"I don't like priests." Tomas responded without emotion.

Nick looked over at his partner to see if he intended the comment as a joke or was serious. Tomas wasn't smiling. "How can you not like priests? They are like... neutral."

"Just don't like them. Never did. You're alone on this one."

Nick just shook his head and sighed.

Saint Francis de Sales was not difficult to find. As Nick arrived in the center of Sao Vincente he saw a small sign for the church and a minute later was pulling into its dirt parking lot.

Nick got out of the car and felt the heat of the sun through his sports jacket. He normally didn't wear the jacket to the office but always had one on hand when he was on the road for interviews. Sometimes it was necessary to look "presentable", and it seemed to earn a little more respect than a simple button-down shirt.

Although outside it was too warm for the jacket today, he was glad that he had it when walking through the front doors of the church. It was much cooler inside. As a kid, he was always told to wear his 'Sunday Best' for mass anyway, so he was still following tradition.

He crossed through the small narthex and in amongst the

pews, looking for someone to inquire as to where he could find the priest. He saw no one except for a lonely worshipper in the front row, an old woman who, with bowed head, was clearly praying. He could hear her rosary beads clicking together as she rubbed them, her small head bobbing up and down in time with her prayers.

Behind him he heard a sound and turned just in time to see a small door close adjacent to the entrance. He hadn't noticed the doorway when he came in, since the lighting in the church was poor, but he now walked towards it in the hopes of finding someone to help him locate Father Angelo.

A sign above the door said 'Rectory'. Bingo.

The door opened to a narrow hallway, but no one was immediately visible. Nick took a step inside and followed the hall to the end, which led to a door that was locked but had a bell.

He could hear the doorbell inside as he rang and very quickly a nun appeared.

"Hello. May I help you son?" she asked.

"I'm sorry to bother you but is Father Angelo available for me to speak with for a few minutes?"

"Confessions are Wednesdays and Saturdays at 10 a.m. and 3:45 p.m., why don't you come back to speak with him then."

"Please. I'm not here for a confession. I'm with the police department." He hesitantly pulled out his badge for proof, but was skeptical of showing any type of authority in situations like this, sometimes it backfired.

"Oh no! What's happened?"

Uh oh.

"What could the police possibly want with Father Angelo?" the nun continued. "Is it his sister? Has she passed away?"

"No, it's nothing to worry about. No one has been hurt or is in trouble. I just want to talk with him about a member of the congregation. It will only take a few minutes."

She breathed a sigh of relief and turned to look behind her in the room. "Please wait here a moment", she said, and then shut the door.

Nick always had trouble dealing with such gentle people. In his line of work he was used to handling cold-blooded criminals, interviewing people of the cloth was a little out of his ballpark. He looked around him and noticed a wooden bench where he took a seat, beside the bench was a small shrine where one could kneel and pray.

When was the last time he had prayed? His grandmother's funeral probably, but that was ten years ago. These days, when he goes to church it's only for formal functions like weddings, funerals, and christenings. As a kid, his parents brought him to church every Sunday, no excuses. He hated getting dressed up in an uncomfortable shirt buttoned up to his chin. He would have to sit quiet for an hour or more listening to an old man talk when all he wanted to do was play outside.

He still believed in God, Heaven and Hell. Maybe he ought to start bringing his boys to church. Whether they liked it or not, they should at least have the chance to make up their own minds about their faith.

As he contemplated his spirituality the door before him opened and a priest dressed in a black robe appeared.

Nick stood to address the small man with perfectly sculpted silver hair who stood before him. "Father Angelo?"

"Yes, Sister Anna tells me you are with the police?"

"That's right. I'm Inspector da Silva with Sao Paulo Homicide and I'm investigating the death of Eduardo Rodriguez."

"Ah, yes, Eduardo." The priest stood with folded hands and bobbed his head as he spoke.

"Is it possible to speak with you about him for a few minutes?"

"Certainly. Follow me please."

Nick followed the priest through the door and into the rectory. They walked by an office and then into the living quarters. His first impression was that it was not what he had expected a priest's home to look like. The living space looked just like a regular home. They entered a sitting room which had several easy chairs and a couch, in one corner was a rather large television and beside it was a stereo system. Adjacent to the sitting room was the kitchen and, as they entered, Father Angelo motioned for Nick to take a seat on one of the benches set at a long table.

Father Angelo walked to the refrigerator and took out a plate of sandwiches and brought them to the table.

"Would you like tea or water with your lunch?" he asked.

"Oh, Father, I didn't mean to intrude upon your lunch. Please enjoy your meal while we talk." Nick had wondered why they weren't conducting this conversation in the office, now he knew why.

"Have you eaten lunch yet?"

"No, but—"

"Then would you like tea or water with your lunch?" Father Angelo repeated his question but it sounded more like an order the second time around.

"Water would be great." Nick forced a smile but was feeling as though he was losing control of the conversation. Trying to get back on track he added, "Can you recall the last time that you spoke with or saw Eduardo Rodriguez?"

"Yes."

Nick waited a moment for the priest to continue but he was apparently finished with his response. Nick didn't like how this interview had begun. With abrupt answers like that, this was not going to be easy. "Could you please tell me about that meeting?"

"No."

"No? Why not?"

"Because it was a confession. I can't reveal our conversation to you."

"Can you at least tell me when it occurred?"

"Last Wednesday afternoon." Father Angelo set a glass of water in front of himself and one in front of Nick, then sat down. "It was one of the final days of his life."

"Yes, I'm aware of that." Nick replied. "Listen Father, I—"

Father Angelo interrupted him again. "No, not yet. I am hungry. First eat and then we will talk."

Seeing no other recourse, Nick gave in and took a sandwich from the tray. It was smoked ham with butter, not his favorite. But it looked like he had no other choice, if he wanted to get some information he would have to eat first.

He dared not speak for fear of being interrupted yet again so he waited and watched the priest sitting across the table from him. The man was quite old and, despite his perfect coif, rather elfish looking with a tuft of hair sprouting from each ear. He certainly was not someone he felt he could reveal his soul to.

He never could understand the calling to priesthood, but after meeting this man he began to see why someone could be happy with that choice. Even a meek man could wield considerable power dressed in the garb of the Holy Church.

Father Angelo nibbled on his sandwich, taking very small bites and a sip of water between each one. After what seemed like an eternity to Nick, the priest took his napkin from his lap and cleaned the edges of his mouth. He sat up straight and crossed his legs, then began talking.

"I'm not sure why you're here, and I don't think there is much information that I can provide you with because most of my personal conversations with Eduardo Rodriguez were in confessions. But what I can tell you is that he was an honest, caring and trusting man, he loved his family more than life, as he also loved God, and he now rests in Heaven with our Lord."

The priest then bowed his head.

Nick wasn't sure if he should say Amen, bow his head in prayer or start applauding. Was he wasting his time? He was afraid that every one of his questions would be dead-ends if the Professor's recent visits had all been confessions. He sat there silent for a moment and then realized that he needed to get to the point. He required some basic information as quickly as possible.

"Father, let me be perfectly blunt, did Eduardo Rodriguez know that he was going to die?"

"We all die, my son."

Nick was at his boiling point. He had tried to remain calm

and to salvage any chance of getting some useful information from this man, but he was at his limit.

One last try.

"Perhaps I wasn't quite clear," Nick said. "Did he know that he was going to die *last week*?"

"Why is this information important to you? What gain is there in subjecting a peaceful soul to restlessness?"

The straw that broke the camel's back.

And half an hour ago he was seriously reconsidering his faith.

"Let me be frank," Nick began. "Mr. Rodriguez died of a heart attack in the early hours of last Friday morning, but normally when people die they leave a few items in their 'In Bin' if you know what I mean. This man didn't. This man wrapped things up like he was going on a permanent vacation, which was apparently the case. I'm not implying that the man was dishonest or unfaithful, all I want to know is the truth. Did Eduardo Rodriguez tell you that he was going to die?"

"I'm sorry. You'll have to go now."

"What?!?" Nick threw his hands up in the air. "I certainly didn't mean to offend you." He was back-peddling now. "Can't you tell me anything at all? I'm trying to help his family."

"His family is my concern, Inspector." The priest stood.

"But—"

The priest held up his hand for Nick to be quiet. "I will see you to the door."

On top of not being allowed to speak, Nick was speechless.

Father Angelo motioned for Nick to follow him into the

main hallway of the house which opened to a foyer and the front door of the rectory.

As Father Angelo held the door open for Nick he made one final remark, "All I can add, Inspector, is that I was expecting the call from Rosalinda Rodriguez on Friday morning." He smiled for the first time. "Have a pleasant day."

With that the priest shut the front door, and Nick heard the finality of his visit when the dead bolt locked behind him.

Nick climbed into the car to find Tomas reclined in his seat with his cap tilted down over his eyes.

"And?" Tomas asked without changing position.

"Damn priests."

Tomas sat up. "Told you."

Chapter Seven

"Boss, you got a second?" Nick asked, leaning a hand against the door frame of the chief's office. Another police officer was in the middle of a conversation with Esposito.

"Sure, come on in Nick," Esposito said, then turned to the other officer and quickly finished their conversation. "I want to see a copy of that report on my desk this afternoon."

"Yes, sir." The rookie officer then hurried out of the office.

Esposito shook his head and stood peering for a moment out the window. Without turning around he said, "You know Nick, times have changed."

Nick was about to make a wise-ass comment but, when his boss turned around with a sullen look on his face, he kept it to himself.

"It used to be that guys went into the force because their fathers were cops and they had grown up with respect for the institution. Now all they want is to have some kind of power over the common people." He turned around and glared straight into Nick's face. "If you're not a cop for the right reasons then you'll never do the job right, and we just don't need that kind of guy working with us."

"Whoa, boss, I hope you're not insinuating..." Nick said, as a guilty twinge ran down his spine.

"No, of course I'm not talking about you. Sorry. Sit down."

Nick took a seat and Esposito got back to business. "What do you have for me?"

"The math professor." Then realizing that his boss wasn't really focused on him, he added. "You okay boss? What was all that about?"

"Nothing." Esposito answered. "Go ahead. The professor. You got that case wrapped up?"

"Wrapped up? Boss, I've got half a dozen open cases. If I dropped them all to take care of this one I think you'd have my badge."

"Yes, I probably would if we weren't so short-handed." Esposito said with a trace of a smile, but still looked as though his thoughts were somewhere else.

Nick hesitated to see if his boss was going to continue with his monologue, but the silence lasted more than a moment, so he began with the briefing. "The family appears to be in the clear, at least the women. They're stunned at the passing, no one seems to have had knowledge of anything suspicious and no indication of ill health." Nick paused to see if he had his boss' attention.

Esposito was right with him. "What do you mean 'at least the women'? Who else is there?"

"There is a son. Mid-thirties. Lives in Brasilia. Tough to read."

"Okay, that's one lead to follow up on. What else?"

"The autopsy report said that cause of death was massive coronary failure brought on by a rare, and essentially undetectable, heart defect."

"*Essentially undetectable*? What does that mean?"

"Simply that there are currently no tests which can detect this defect, and because of its rare occurrence, it isn't looked for even when heart problems are evident. At least that's what his doctor told me."

"Check. What else?"

"No other wounds to the body, no sign of drugs of any kind."

"Then maybe it's a closed case after all."

"Not quite." A wry smile came to Nick's face. "I can quote you as saying that this was 'an interesting case', and you know what? You were right."

"Just give it to me. What else you got?"

"I stopped by the funeral home that handled the burial. It turns out that they had an *arrangement* to handle Mr. Rodriguez. Everything was ordered and paid for two weeks *prior* to his death."

Esposito cocked his left eyebrow, which Nick knew meant that he wasn't buying it.

"Seriously. A man telephoned and, without leaving a name, discussed burial costs, etcetera, etcetera, and then sent a courier with specific requirements and cash to open an account in the name of Eduardo Rodriguez."

"Interesting. Did the funeral director or someone there know conclusively that it was the professor who made the arrangements?"

"No."

Esposito smiled, knowing that Nick wouldn't let this valuable information slip through his fingers. "But you got the courier didn't you?"

"Of course I did."

"And?"

"Confirmed the source of his shipment to be a professor at the university, but he didn't have a name."

"And??" Esposito didn't like to have to drag the information out of his subordinate.

"I showed him a photo of Rodriguez." Nick sat calmly with a wry smile.

"And?!?"

"We have a winner!"

Esposito rolled his eyes. "You're too much, Nick."

Nick went on to describe meeting the priest, interviewing the family members and speaking with the lawyer who handled the will.

The family lawyer appeared to have no knowledge of the death prior to the occurrence and said that many people regularly update their will with the addition of new grandchildren or other changes to the family structure.

Esposito sat thinking and then tried to summarize what he had heard. "So, the only people that we know of so far with confirmed prior knowledge of the death were the priest and the funeral home director?"

"And the professor."

"Right, and the professor." Looking across the room with a blank stare while tapping a pencil on the desk was Esposito's way of concentrating. Nick knew it and kept quiet, waiting for his boss to continue. "No drugs found in the autopsy? Medications?"

"Nothing."

"The autopsy was conducted three days after the death?"

"Correct. He died in the early hours of Friday morning and it wasn't completed until Monday morning."

"Why the delay? They normally work all weekend."

"Somebody was on vacation and they were understaffed."

"Okay, so what do we have for traceable drugs which induce heart failure?" Esposito was half talking to himself while still thinking, then rambled off a couple of names of drugs that Nick had never heard of. "The question is, can any of these drugs be absorbed into the body of a corpse and become untraceable after three days?"

"I'll look into it boss. But even if we have a *means*, we still don't have a suspect. Everyone tells me this guy was squeaky clean. Family man. Church go-er. Beloved teacher. I can't find a motive."

"C'mon Nick, be creative. You must have a theory. What is it?"

"Okay, the man was not wealthy, sure he was comfortable, but no one was going to get rich by knocking him off. Since all his money was distributed among the family prior to death, there is no major windfall to be claimed by a single party, and I seriously don't think his whole family conspired to get rid of him. Therefore the only possible motive I can come up with is knowledge-based."

"Knowledge-based? What the hell does that mean?"

"It means that the guy was a professor and a researcher. He was recognized worldwide in his field and may have had something that someone wanted to get their hands on."

"Nick, the guy was a math teacher. What could he possibly have had? A nice calculator?"

"Just listen." Nick continued. "This guy was a genius from

what I hear. Maybe he was working on something that someone else wanted to take credit for. I mean, recognition in one's field can mean a lot more to some people than any amount of money."

"Interesting theory." Esposito smiled. "Any candidates at the university or are we looking at a worldwide field of players?"

"I've got some ideas."

"Okay, follow up on it, but check into the angle with the son too."

"I will." Nick got up to go.

Esposito was still turning things over in his mind. "Nick, something's not right here."

"I know."

"Find out what it is."

"That's my job, boss."

"Heading to the University?" Esposito asked.

"A no-brainer, right?"

"I taught you well." Esposito laughed and, as Nick was heading out the door, he added, "Keep away from the coeds!"

Chapter Eight

Ramon Santos sat across from Professor Montagne and was vigorously explaining the goals of his thesis work. He was very disappointed about having to complete his dissertation without Professor Rodriguez, as he was sure that no one could match the brilliance and exemplary tutorial of his late mentor. Unfortunately, a professor was required to sponsor doctoral research so he had to find a replacement and Professor Montagne was the next obvious choice.

Ramon had worked very closely with Professor Rodriguez for several years, and Ramon looked to him not only as an instructor but also as a father figure, having lost his own father at a young age.

To make matters worse, theses have to be published under the names of both a professor and a student. He felt that it was a bit unfair that Professor Ricardo Montagne would get credit for the research when Professor Rodriguez had been the driving force behind it until now. Perhaps he could use both professors' names on the final draft.

Ramon was also unsure whether Montagne even had the knowledge in this specific field to guide him as a PhD student. His thesis involved modeling elastic properties of a new composite material to be potentially used for spaceflight applications. The work not only required knowledge of advanced math, but also an in-depth understanding of the theories of materials engineering. Montagne's primary field of study was in growth and decay, and he had little or no experience in experimental work, which was a large part of Ramon's research.

But, as his mother always says, "Everything happens for a reason."

He wasn't sure what the reason was for the death of his mentor, and he couldn't fathom that it was a fair one, but he decided that he needed to make the best of the situation and put all of his efforts and concentration into his goals, despite the traumatic set-back.

And in order to do that, he needed to have a good working relationship with Professor Montagne.

"Yes, I understand your objectives," said Ricardo, "and I look forward to continuing this course of study with you." Ricardo stood up and walked around to the front of his desk while leafing through his mail. "Why don't we go down to the lab now and you can show me your experimental setup."

Walking next to his newest research student in silence, Ricardo was disappointed in the topic of study that he was pursuing. Granted he felt capable of reviewing the advances in the field enough to guide Ramon through it, but he was hoping that the topic would be more along the lines of his own so that he could acquire some insight from the young, promising mathematician.

He should have expected as much. Professor Rodriguez had always been involved with topics that focused on new technologies, combining his mathematical knowledge with engineering principles. He had shown little interest in Montagne's growth analyses. Sometimes Montagne even got the feeling that his colleague had little respect for conventional math.

But perhaps the untimely death of the esteemed professor was a blessing in disguise for him. Now he would receive credit for investigative studies in this new field and will essentially be expanding his area of expertise at the same time. He will certainly be the most respected and sought after mathematician in Sao Paulo, and possibly in all of Brazil. He languished in the thought of endless possibilities.

As they started down the stairs toward the laboratories they crossed paths with Professor Kumar, who had another graduate student in tow.

"Ah, Professor Montagne, I see you have already begun advising our experimental modeler. Maybe you will learn something too!" Kumar chuckled to himself. "Actually, could I have a word with you?" Then turning to his student he added, "I'll be right with you." Walking a few steps away, he gestured for Ricardo to follow.

The two graduate students began making small-talk while waiting for their respective professors to finish their side-bar.

"What is it Amar?" Ricardo asked with a sigh.

"Ricardo, we talked about you taking over Eduardo's class schedule and graduate student research for the remainder of the year, at least until we can find a replacement—"

"Yes, of course, I've already spoken with all the graduate students. It's no problem."

"Well, I have another request. I know that you have limited

time on your hands and this may take some considerable effort, but would you mind sorting out Eduardo's office?"

Sorting out his office? Was he the custodial help now too?

Ricardo took a deep breath and turned his head away. Of course he was the logical choice for this task considering that he would be taking over much of the professor's work, but it was certainly not going to be pleasant sorting through a deceased colleagues work. He tried to deflect the blow. "Don't you think that you should take care of it yourself, as the Department Head?"

Amar nodded as though he had already thought it through. "I really think you ought to do it Ricardo. You are overseeing his students and classes, and you would know better than I what papers and files should be kept and which could be discarded." He then lowered his voice. "Besides, I don't want to go rummaging through a dead man's belongings! Creepy!" Amar laughed at his own little mock scared routine then motioned for his student to follow him upstairs. As he paced backwards he added, "Stop by on your way back up and I'll give you the key to Rodriguez' office."

Damn it. That's going to be a lot of work. But as Amar had mentioned, he was the only one who would know which files should be kept with the department, which ones he would need for himself to continue the work with the graduate students, and which could go in the trash. He sighed deeply, then turned to Ramon and said, "Okay, let's get back to work."

"So, Professor Kumar wants you to clean out Professor Rodriguez' office?" Ramon asked.

"Yes. He likely has class notes and research files that need to be retained by the department. I'll have to go through everything with a fine-toothed comb and see that each finds its proper place."

"I'd be happy to help you go through the files. I know it's a lot of work, but I feel as though I owe it to Professor Rodriguez for all the help he's given me. Besides I've spent a lot of time working with him and I may be able to identify some items that you can't."

"I appreciate your offer Ramon, I'm sure I won't have any trouble *identifying* the files, but you may be able to help me speed up the process a bit." Ricardo looked at his watch. "I really don't have much time during the day so we'll have to do it after hours if that's all right with you."

"Sure, anytime," Ramon responded as they entered the laboratory.

"Then, how about meeting at five o'clock this afternoon in his office?"

"Alright. I'll be there."

"Good." Ricardo was actually feeling a little relieved by the offer of help. "Now, Ramon, why don't you show me what you are doing with these materials and how your experimental and theoretical analyses are matching up."

The hallway was deserted as Ricardo put the key into the lock of Eduardo's office. He paused for a moment before turning it. Eduardo was the last person to be in the office prior to him, and now he was dead, a week ago yesterday was the last time he was here. He groaned inwardly as he turned the key and pushed open the door.

To his surprise there were no foreboding shadows drawn across the office, nothing to imply that its previous resident would not someday soon be returning to resume work in its space. He shut the door behind him as he did not want curious passers-by to happen in on him. Then he looked around, trying to get a feel for where he should begin.

It was odd being in this office alone, he felt like a trespasser, which certainly he was. Even though most of the contents of the room were property of the University, he knew from his own experience how private an office and personal files are. He would not want anyone sorting through his own office, but death is the end to all, public identity and private belongings alike.

There was a bookshelf to his left that rose from the floor to the ceiling, it was full of binders with labels ranging from course numbers to graduate student research topics to published papers from journals around the globe. Another wall of shelves to his right was lined with books, next to that was a set of filing cabinets and before him stood a desk with two chairs in front and one behind—Eduardo's chair.

Perpendicular to the writing desk, on a small side table, sat the computer. Sitting at the computer, one had the vantage point, when looking slightly to the right, of a view out the window and onto the campus quadrangle.

Ricardo walked behind the desk and peered outside. It was essentially the same view as from his office, just a little farther east and sheltered by a few trees, which helped filter the hot afternoon sun. Perhaps he should move into this office. Maybe he could even gain some inspiration within these four walls, as its predecessor had.

Just then there was a knock on the door and Ricardo

turned to see Ramon peeking his head inside. "Are you ready to get started Professor Montagne?"

"You know what Ramon? I think I'll need a little time to clear out Professor Rodriguez's personal effects and get a feel for where everything is before we can sit down together and start sorting things through."

"Okay, should I come back later?"

"Humph." Ricardo looked at the magnitude of the endeavor surrounding him. "It may take me a little time. Besides, I'm sure you don't want to be hanging around here, digging through old paperwork on a Friday night. Why don't we meet here again on Monday, at the same time?"

"All right, if you think so, but I really don't mind staying if I could—"

Ricardo cut him short. "No, go ahead Ramon. I think I'll be spending a good portion of the weekend in here. But we can meet again on Monday and then you can help me put everything in its rightful place."

"Okay, if you're sure." Then as almost as an afterthought while taking off down the hall he added, "Have a nice night Professor!"

"You too." Ricardo mumbled even though he was sure that Ramon was already out of earshot.

Have a nice night. Ramon's words echoed in his ear. From the looks of the office, he figured it would be a long one.

When was the last time that he had plans in the evening? He couldn't recall. All his social engagements seemed to be associated with the University. Once in a while he would venture alone to the opera, but he had trouble recalling the last one he'd seen. It was a depressing thought knowing that his work would likely be his only true life companion.

He sat down in Eduardo's chair and noticed the small group of framed photographs in one corner of the desk. A shot of him and his wife standing on a pier with the sun setting behind them; another was a portrait of his son, Miguel, and his family; and then a third photo of his wife and daughter, Maria, looking at each other laughing. A handsome family. A pity for them to lose the head of it so unexpectedly.

Another knock on the door shook him from his reverie. "Yes, Ramon?" he inquired, assuming his student was back with a question.

The door opened but it was not Ramon who stood there. The face was familiar but Ricardo couldn't quite place it at first.

"Good afternoon, Professor Montagne."

Realizing that he wasn't recognized, Nick added, "I'm Inspector da Silva, we met at the Rodriguez home on Tuesday."

"Oh, of course, Inspector." Ricardo stood and shook Nick's hand. "Please take a seat."

Nick sat in one of the chairs placed in front of the desk, normally used by the students.

"That's quite ironic," Ricardo continued. "I was just now thinking about Professor Rodriguez's death. Then you arrived."

"I'm not sure I see the irony. I mean, you are sitting in the office of the deceased, are you not?"

"True, true, but I'm not here in mourning, rather I'm trying to salvage relevant material for the department."

"It doesn't look like you've gotten very far," Nick said while looking around at the orderly office.

"No, I've just begun." Ricardo didn't like the implication that he wasn't making headway. "And I have a lot to do, so why don't you tell me why you are here."

Ricardo wasn't comfortable with the man who sat confidently before him. Inspector da Silva made him feel out of sorts, and of all places, where he should feel most secure: the University. He wanted the inspector gone as soon as possible.

"As you already know, I'm investigating the circumstances surrounding the professor's death, so naturally I wanted to examine his office space." Nick unbuttoned the sleeves of his shirt and proceeded to roll them up to his elbow. "I left a voicemail with Professor Kumar this afternoon that I would be by. Didn't he relay that message to you?"

"No, he did not." Ricardo wondered why he always seemed to be the last to be informed of anything. But in defense of his colleague he added, "But I'm not surprised. He must have thought it was routine for you to come by." Then Ricardo sat up straight, leaned slightly forward, and looked the Inspector directly in the eyes. "Although, it is difficult to imagine that foul play could have been involved in Eduardo's death."

"Why is that?"

"Because the man was practically a saint!" Ricardo said.

"A saint?" Nick jumped right on it.

Then Ricardo realized that his statement may have revealed some of his true feelings towards Eduardo, so he tried to clarify, "I mean—".

"What do you mean, Professor?"

Ricardo sighed deeply and in an attempt to show genuine

emotion, looked at Eduardo's family photos and shook his head before continuing. "Yes, a saint. He was brilliant with the students, respected by colleagues, and renowned in the world of mathematics....like a saint, he was loved by all."

"Except you."

Ricardo was caught off-guard by the Inspector's remark and looked at him with both surprise and anger. Looking for clarification in the hopes that he had not heard correctly, he said, "I'm sorry?"

"Everyone loved him, except for you. He was your arch-rival, wasn't he?"

Ricardo got his emotions in check and paused to organize his thoughts. He couldn't believe that he may now somehow be caught in the middle of some wild goose chase by an over-confident police Inspector concerning Eduardo's death.

And who did this Inspector think that he was, talking to him that way? He was a respected University professor, not some street thug. "Inspector, please, if you have business that you need to attend to here, then by all means do it, otherwise I have little time for accusations and insults."

"My apologies, I had no intention of insulting you," Nick smiled, not out of humility, but rather from enjoyment. He liked the power that his position provided him. He could say almost anything, no matter how insulting, and still get respect from people. Well, except for priests.

Then a sense of déjà vu reminded him of his boss's comments earlier that afternoon, about how cops today liked to display power over common people. Maybe he should in fact change his tactics, try to be a better man.

But, looking at the man across from him, he felt that there

was more to the professor than met the eye, and he decided to pursue his offensive approach. "I would like to have a look around the office, dust for fingerprints, you know, detective-type stuff."

Ricardo jumped up and huffed like a raging bull.

Nick enjoyed the reaction.

"How long will this take?" Ricardo asked. "I have a lot of work that needs to be done in here."

"Oh, a couple of hours should be all I need, then it's all yours."

"Fine. I'll be down the hall in my office. Let me know when you're done." Ricardo started for the door. "Now it looks as though I'll have to come in over the weekend to get my work done in here."

Nick figured the professor probably spent most of his weekends in the office anyways. "Oh, and one more thing Professor."

"What is it?" Ricardo asked with pursed lips.

"I'll have to ask that all contents remain in this room until my investigation is concluded. Any relevant material that you should require shall only be photocopied and then returned to its original location."

"Are you serious? This is ridiculous! The man died of a heart attack in his sleep!"

"*Allegedly*," Nick corrected. "Unless you know something that I don't?"

With that Ricardo had had enough and, accompanied by an angry growl, he stormed out of the office.

Nick smiled to himself as he walked around the desk to sit in the Professor's chair. He didn't plan on dusting for prints.

He knew that he'd find half the staff and dozens of students had been in there and it wasn't critical to his investigation. But he had to give some kind of official grounds for being there, especially since he had lied about calling Professor Kumar.

The element of surprise can sometimes yield considerable useful information in a case.

Reaching over to turn on the computer, Nick noticed the family photos on the desk. Such a shame for the wife and daughter, but the son, Miguel, may be another story. Nick wasn't so sure that he was accurately portrayed as an innocent victim.

Time would tell.

He booted up the computer and was happy, albeit surprised, to find that no password was required. Wow, the professor was very trusting. Clicking on the email icon was his first move. He searched through the Inbox, Sent Mail, and Deleted Messages but found nothing out of the ordinary.

He then went to the C:/ drive and looked under all the file folder headings: MA101, MA402, MA403, and so on. He opened the folder marked MA101 and noticed it was filled with student rosters, grades, and some notes. He assumed the others would be the same so he began to browse through some of the other folders.

He came across information on conferences, publications in the field of mathematics, and a multitude of words that he had never heard of and thus assumed to be mathematical jargon.

Nothing so far. It seemed like a dead end.

Nick scratched his head and tried to think where he might

find something personal on this man. Although, in all likelihood, he probably didn't keep anything personal in his office space other than the photos on his desk. There were no personal files, no letters of any sort, nothing other than work-related topics.

There were other software programs loaded onto the computer that he didn't recognize, but he didn't have the knowledge to try to locate files on them that would potentially be of any use to him. If it were to become necessary later, then he could have a police expert come in and sort through them.

Nick's knowledge of computers was relatively limited, but as a last effort to find something of value, he checked the Documents folder under the start icon.

And that's where he came across something interesting.

The folder appeared to have been recently cleared.

But one file remained.

The title: *Upon_My_Death.doc*

Chapter Nine

The telephone rang and Sophie yelled to her mother that she would get it.

"Hello. Sophie Stanley speaking."

"Hello Sophie, this is Doctor Mansfield, could I speak with your mother please?"

"MOM!!!! It's Doctor Mansfield!!!" Sophie dropped the phone on the table and ran back to her game.

Caroline picked up the phone as it teetered precariously on the table, "Hi, Doctor Mansfield. Sorry about that."

"No need to apologize. I think it's so cute when kids answer the phone, and she was awfully polite."

"Until she broke your eardrum with her screaming."

"Don't worry. My ears are just fine." Dr. Mansfield paused for a moment before continuing. "Caroline, I have the test results from Ron and your sister. They are rather interesting."

"Interesting? Well, that sounds encouraging." After a moment of thought she added, "But not promising. You mean neither were a match?"

"Unfortunately, no, neither were a match. But your husband seems to actually be a partial match. He has two HLA molecules which are identical to yours. The chances of this occurring between two married people is minuscule, unless they were related of course." She let out a short laugh. Then upon thinking about her statement, she added seriously. "You and Ron aren't related, are you?"

Now it was Caroline's turn to laugh. "No, of course not. Only by marriage."

"Good. So, I've talked with a colleague of mine who is a specialist in leukemia treatment and has dealt with thousands of patients. He said that it was completely unheard of to have the spouse be a half match."

"Does that mean that he can be my donor?" Caroline asked with enthusiasm.

"Well, no, but what is does mean is that I'd like to test Sophie now. If by chance she has inherited the two HLA molecules from Ron that match yours, then she will be a perfect match. The chances are good, one in four to be precise."

"My goodness. Well, I guess that's good news, although I do have mixed feelings about it, considering that Sophie is so young."

"Not to worry Caroline, as I told you before, there is really nothing at all to the test, just a small blood sample, and if by luck she is a match, then I guarantee you the donation will go smoothly." Then she added, "Right now, she is the best chance we have."

"Well, I'll have to talk to Ron about this. I mean... I can bring Sophie in for the blood test, but, well... I guess I just need a little time to let this digest."

"I completely understand. Let's get her tested and then we'll talk further, okay?"

"Okay."

"But let's not wait too long. Can you come in on Monday morning, right at 8 a.m.?"

Realizing time was crucial, Caroline answered, "Okay, Monday morning then. And... Doctor Mansfield..."

"Yes?"

"Thank you." It felt good to have someone making the decisions for her.

"No need to thank me now, wait until you are in remission and then you can thank me!"

"All right, I'll wait until then," Caroline said with positivity.

Caroline hung up the phone and walked into the kitchen. Through the sliding glass doors she could see that Ron and Sophie were now in the backyard playing catch.

It was another unusually mild day, perhaps the typical March winds would spare them this year. But that was probably wishful thinking, Caroline thought.

She walked out onto the deck, and Ron saw her just as he caught the ball. He then coaxed Sophie into checking on the rabbits to see if they had enough food and water before approaching his wife.

Caroline began, "That was Doctor Mansfield on the phone."

"I know. Did she have good news for us?" He reached for her hand as he spoke.

She shook her head, reminded that Ron was not a donor match, but then she added, "Well, yes and no."

"What does that mean?"

Caroline then told him about their conversation and the rare occurrence of their genetic similarities and how their children have a one in four chance of having identical genetic matches.

"That's great news, honey! That's incredible. See? I knew we were going to beat this!"

But Caroline couldn't share her husband's enthusiasm and Ron seemed to sense it.

"Caroline, please, we have to be realistic here. Doctor Mansfield told us that the procedure is not very involved and that Sophie would recover quickly. We can't even think twice about this if your life is at stake."

His words hit her like a speeding train. *If your life is at stake.*

They hadn't really talked about death. They both assumed they would somehow, some way, find a cure. But now it hit her, after hearing Ron's words, that she could actually die.

Ron could read his wife like a book. "Come here," he said as he guided her to a patio chair and took the one next to it for himself. "Honey, I know that we really haven't talked about what may happen if we don't find a match, but I think we both know what the reality of the situation is. I propose that we don't focus on the worst case scenario unless we have to. Right now, let's concentrate on getting you cured, and in the meantime, we live our lives as normally as we possibly can. It's the beginning of spring, and with this beautiful new deck we are going to have a great summer filled with friends, family and barbecues."

Caroline's lower lip was quivering as though she were about to cry, but she took in a deep breath and slowly let it out. "I guess you're right, for now anyway." She saw Sophie running towards them. "God, I love that child so much," she whispered.

"And she loves you. She'd do anything to help her mommy."

Sophie hopped up onto her mother's lap and announced that Bunny Rabbit was eating and that Br'er Rabbit was sleeping on his back.

"On his back???" Ron asked, slightly alarmed.

"Yep! He's sound asleep, I even rubbed his belly and he didn't wake up!"

Ron and Caroline gave each other a knowing glance.

Ron walked out to the cage while Sophie and Caroline played Paddy Cake. Sure enough, one of the rabbits was dead. He took a shovel out of the work shed and dug a deep hole at the back edge of the yard. He scooped the dead rabbit from the cage and carefully laid it to rest, piling the loose dirt on top and then placing a large stone upon it to mark the spot and keep predators from digging it up.

Twenty minutes later, as he was climbing the steps back onto the deck, Sophie asked, "Is Br'er Rabbit awake now?"

Ron took his daughter onto his lap. "No sweetheart, Br'er Rabbit isn't awake, but he's not sleeping either. He went to Heaven."

"Where's Heaven? And when is he coming back?"

"Heaven is where great-grandma is, remember? And once someone goes to Heaven they don't come back."

"Oh yeah, you told me that already. And we can't visit them either, can we?"

"No, we can't visit them. But someday we will go to Heaven too, and we'll all be together."

"Yea! We'll have a big party when we all get there." Sophie turned back to the empty cage. "But now Bunny Rabbit is all alone." She thought for a moment, twirling a strand of hair through her fingers. "Maybe we should get a cat."

Caroline laughed. "Sophie, we don't live on a farm. Besides, we have Max, and cats don't get along too well with dogs. Why don't we get a new baby rabbit to keep Bunny company instead?"

"Okay, but when Max goes to Heaven, can we get a cat?"

"Sophie! Hopefully we'll have Max here with us for a while! Dogs live a lot longer than rabbits."

"Come here, Max!" Sophie shouted to the poor dog sleeping next to the house.

Max lifted his head to see what was going on. Upon seeing it was only 'Sophie the Dog Torturer' looking for him, he let his head drop with a thud to continue his slumber.

Lifting his daughter from his lap, Ron tossed Sophie in the air a couple of times yielding loud squeals of joy from the little girl.

Then, taking a break to catch their breaths, Ron said, "Sophie, mommy is going to take you to the doctor Monday morning."

"But I'm not sick!" She interjected with a pout before he could finish.

"I know, I know, it's not because you are sick. The doctor just wants to take a little blood from you to run some tests. It will only take a few minutes. And if you promise me you'll be a good girl at the doctor's, when I get home from work we can all drive over to the Riley Farm and get a new rabbit. How does that sound?"

"Yippee!" Then squirming herself loose from her father's hold, she skipped her way across the deck and down the steps. Calling back she said, "I'm going to go tell Bunny Rabbit that he's getting a new friend!"

"So, that was easy enough. Let's keep our fingers crossed." Ron squeezed his wife's hand and then picked up his iced tea and took a long drink.

"Hey, did you say you have to go into work tomorrow? On Saturday?"

Practically choking on his tea, "Oh, that's right, we haven't even talked about it with all else that's going on." Ron tried to catch his breath after inhaling half the drink. "Remember? I'm traveling to the International Architecture Symposium next week. And that's why I want to go in to the office tomorrow for a bit, to prepare our presentation material."

"That's right, you mentioned that a while back. I'd forgotten. You had to miss the symposium last year because of the Jones Tower contract, you couldn't get away."

"That's right. But last year the conference was in San Francisco, where I've already been a couple of times, this year it's in Brasilia!"

"Oh, wow. I remember now. I'm so jealous." Then, as an afterthought, "But is it safe to travel there?"

Ron chuckled. "Yes, honey, it's safe. Do you think they would have architects coming from all over the world if it weren't? Besides, Brasilia is an architect's dream city, although I've heard it's not very pleasant to live there."

"Why not? Crime?"

"No, honey, relax." Ron shook his head at his wife with a

teasing grin, knowing that it was only her "family protection instinct" acting up again.

The history of Brasilia intrigued him, so he decided to share it with his wife. "Actually the city of Brasilia didn't even exist until the 1950s when President Kubitschek decided that he wanted to build a new inland capital city. He hired an architect, an urban planner and a landscape architect who put together the designs for what was supposed to be the world's greatest planned city, with a massive man-made lake and all. It was built in only three years' time and the country's capital was moved from Rio to Brasilia in 1960."

"Sounds nice being able to move to a brand new city. So why do you say that it isn't is a nice city to live?"

"Well, apparently the city covers huge distances and isn't very suitable for walking because there are hardly any trees to protect people from the hot sun, not to mention the lack of greenery for aesthetic appearance. Also, it's supposedly very sterile and organized, not someplace people like to call home. Almost everyone who can afford to leaves town on the weekends."

"Hmmm, I guess an inland city in that climate may not be so nice. But that's easy for me to say since I couldn't ever imagine not living near the ocean. But it sounds like the perfect place for an architectural conference."

"That's for sure. I remember studying the planning of the city in college, but I never dreamed that I'd actually get a chance to go there."

Then Ron wondered whether leaving home at this time was such a good idea. "Do you think I should stay home? I mean, if Sophie has to go through with this procedure, I need to be here."

"No, we need to keep our lives as normal as possible, just as you said." Caroline knew how much this meant to Ron and there was no way she would deny him this opportunity. "Besides, I want you to go, and we don't even know if she'll be a match! So let's wait until we have to make some decisions before we start changing our lives."

Ron smiled at his wife thinking how much he loved her. "You are amazing."

"So are you," she said while poking him in the gut. "I'm going to get dinner ready."

Caroline stood and looked to the backyard and the rabbit cages. She called to her daughter. "Sophie! Can you help me set the table please?!?"

Sophie came skipping up to the deck to follow her mother inside, but Ron grabbed her first and tickled her, despite her giggles and squeals, before letting her go on with her chores.

Later that night, as Caroline was putting Sophie to bed, Ron went to the home office computer and logged on. Once the homepage finished loading, he typed 'leukemia' in the search field and clicked enter. This was about the tenth time he had made such similar queries over the past week. The amount of information on leukemia was overwhelming.

He had already searched for research institutes, experts in the field, medications and methods of treatment, but this time he was looking for something else: Life expectancy in varying stages of diagnosis.

Certainly he knew that he would find no definite answers; he understood that the doctors could only give a range. They had not yet broached that question intensely with Doctor Mansfield, but he was hoping that he could find some personal stories that would shed a different light on their situation, less clinical, more intimate.

He wanted to know what he was dealing with. Months? Years? Perhaps a decade? New treatments were always being found, and maybe if Caroline could survive the next several years, a cure will eventually be discovered.

He had spoken on the phone that week with several experts including a physician who had been working with leukemia patients for twenty-five years. The doctor said that when he first began diagnosing patients with leukemia, he immediately informed them of the dire outlook, that is, patients in acute stages were left with little hope. Today, everything is different he said, there had been so many new drug treatments found, not to mention earlier detection screenings, which lengthened and even saved the lives of thousands of patients each year. He was optimistic that within his lifetime, he would see an end to this dreadful disease.

The doctor even knew of Dr. Mansfield and had spoken with her on several occasions. Ron went on to explain Caroline's case to him. He said that it sounded very typical and if a donor were to be found that she would have a good chance of full remission. He then informed Ron of the rare chances of finding an exact match and the problems associated with partial match donations.

Ron had heard all of this before from Dr. Mansfield but it was reassuring to hear a field expert with considerable clinical experience say the exact same thing that Caroline's doctor had told them.

After the familiar list of websites appeared under his query, he chose one that he hadn't been to before and, with a few clicks through the maze of internet sites, Ron came across a page which posted personal stories entitled *Triumph and Sorrow*.

He clicked the first one on the list under the name *Vanessa Pedrosa* and read her story:

I was training to run the Boston Marathon. I had run it three years before and began with the same training regimen of two 5-mile runs during the week and a 10-15 mile run on the weekend, but soon I couldn't make it any further than the base 5 miles; I was exhausted. I thought that I was coming down with the flu or something, but a simple blood test at the doctor led to further tests. And two weeks later I was confirmed with acute lymphocytic leukemia. I don't understand it. I'm only thirty-four years old, not to mention active and healthy... or so I thought.

Ron read on about her experience with chemotherapy and her search for a bone marrow donor, which was never found. The final paragraph was written by her sister.

...Vanessa died ten months after her diagnosis was confirmed. In less than a year she had deteriorated from a vibrant, happy human being to a weak, drug-dependent frail woman who looked twice her age.

Ron shuddered. Less than a year? My God, could that really happen to Caroline?

"What are you reading?" Caroline had entered the room and was approaching Ron as he quickly closed the page he was on. He wanted to protect her as much as he could despite the fact that he knew he was up against something much more powerful than himself.

"I'm just surfing... looking for options for us."

Of course he knew that Caroline was fully aware of what he was doing, as he had been sitting at the computer until late into the evening almost every night that week. He felt helpless and his thorough internet searching was an outlet for him. Besides, for all they knew, it could prove beneficial to them, but if not, at least he was getting well informed and would feel that he was doing all that he could.

"Ron?"

He turned to his wife and saw the serious look on her face. He knew a heavy conversation was on its way.

"Can we talk about this? I mean, *seriously* talk?"

Ron took a deep breath. "As much as I don't want to, I guess we need to. Why don't you start."

She sat down on the loveseat and he joined her.

"I might die," Caroline said.

Ron tried to speak up but Caroline held up her forefinger to his mouth. "Let me finish." She hesitated for a moment. "I know that we will do all that we can to beat this, but we also have to be realistic and plan for the worst. I need to feel that if I'm unable to take care of Sophie and you, that you will both be okay."

Again Ron wanted to speak but Caroline continued. "According to what I've read about this cancer, many patients are not able to find a donor and die waiting. I figure that if I can't find a donor I have anywhere between six months and a couple of years."

A tear fell from Ron's eye as he listened to his beloved wife speak. "I don't want to die, but if that's what God has planned for me then I have to accept it...**we** have to accept it."

"Not without a fight."

"Ron, I know this will not be easy, and I know that it will be the most difficult on you if you have to raise Sophie alone. She is young and will be able to get through this, but if the time comes that we need to talk to her about her mommy joining Br'er Rabbit and our grandparents in Heaven, then we need to handle this properly. We need to let her know how much she is loved and that she will always be taken care of."

"Of course, honey, I totally agree. But we need to prepare ourselves first before we can prepare her." He hesitated a moment. "Caroline, I can't bear the thought of you leaving us... and I'm not sure that I can even talk in those terms unless we know for sure that that is what we have to deal with. I think that we should wait until we have more definitive answers about your condition and prognosis before we start talking about finalities."

Caroline sighed deeply. She stood up, reached for the mouse, and began shutting down the computer. "I agree with you in some respects. I mean, there is no need to increase our burden if it's not going to be necessary. But I just had to get that off my chest."

She then sat down on her husband's lap. "Let's wait until we get the results from Sophie's blood test and then we'll talk to Doctor Mansfield, in depth, about our options. But I just wanted to open the subject with you because we may have to make some very difficult decisions sooner or later."

"Hopefully later, if ever," Ron said.

"Hopefully. But, in the meantime," Caroline said, "you are coming to bed." She stood up and pulled him up along with her.

"I guess you are right, and after being up late every night this week, I could use a good night's sleep."

Without turning around Caroline responded, "Who said anything about sleep?"

Chapter Ten

He double-clicked on *Upon_My_Death.doc* and held his breath.

Moments later a small window box appeared in the middle of the screen. *Enter Password*, it read with a blinking cursor waiting for input.

"Damn it." Nick cursed out loud at the computer. "So close and yet so far." He clicked on it a couple of more times out of frustration but nothing was apparently going to happen without that password.

He turned off the main switch and unplugged the monitor, he then pulled the terminal plug from the wall and dragged the unit out from underneath the desk. "I guess you'll be coming with me," he said resignedly while patting the top of the metal box like it was the head of a puppy.

He then began searching the Professor's desk drawers: pencils, paper, hole punch, stapler, calculator, nothing interesting other than a funny little gremlin-like statue that was inscribed with 'World's Greatest Teacher'. He slid the drawer shut with a sigh.

He turned around and noticed the nice view of the quad through the office window. It was Friday night and there was a good crowd of students listening to a band playing in the far corner. He drifted back to his own days at the University. He had always tried to give himself a light schedule on Fridays so that he could spend most of the day on the quad. Sonja would meet him after her classes and they would lay there in the sun and talk or play Frisbee, meeting up with other friends who were passing by on their way to and from classes, to the library or dorms.

Looking back now he thought those were easy care-free days, but at the time he felt the weight of the world on him: getting good grades, deciding what career path to take, lining up summer jobs and saving enough money to buy a used car. If only he'd known then how good he had it!

Not that he had a rough life by any means, but supporting a family and the responsibilities of being a parent surely outweigh the stresses of being a student.

Nick decided to look through the filing cabinets next, but was confronted only with pages of notes which, for all he could tell, looked like scribblings of ancient Greek. He recognized a few symbols from the one-and-only calculus class he had taken, but had absolutely no idea what he was looking at. His study of Criminal Science had been light-years away.

After about ten minutes of scanning the files, he had had enough of feeling ignorant to the world of equations and headed over to the bookshelves. More of the same feeling of inadequacy overcame him as he scanned the titles: *Time Series Analysis*, *Formulas for Stress and Strain*, *Probability and Statistics*, *Coding Theory*, plus other titles with words that he had never heard of.

He'd seen enough.

He took one last look around the office, then got on his hands and knees to scan the floor and look under the desk.

Nothing.

Nick deduced that if the Professor was careful enough to clear his computer of all personal files, then it was unlikely that he would find hard copies lying around of anything that would be useful to his investigation.

Well, at least he had a potential lead with the mysterious file. Holding the computer securely under one arm, with electrical cord dangling behind him, he walked out of the office and down the hall.

Professor Montagne had just closed his office door and was turning the key in the lock as he saw the Inspector heading towards him. Not looking forward to another unpleasant encounter he simply waited until the Inspector spoke first.

"I'm finished with the office... for now, but I'll be back, to return the computer at least." Nick patted the top of the unit again.

"You're taking his computer? There's got to be dozens of files on there that could be helpful to the graduate students. What could you possibly want with that? Oh, forget it, don't answer that." Ricardo breathed a sigh of defeat. "The department would like to have the computer back as soon as possible, with all files intact." He wanted this man gone as quickly as possible. "Are you through with everything else in there?"

"Maybe, I'm not sure yet. I'd still like to ask you to leave everything where it is, making only photocopies as absolutely necessary, and returning the originals to where you found them."

Nick glanced at his watch before turning to go and added,

"Well, I guess we'll be seeing each other again soon. Oh, and could you please lock Prof. Rodriguez's office on your way out, that is, if you are privy to a key?"

"Of course I have a key Inspector, I'll be sure to lock it tight." Ricardo tried to play along with the game, but a sense of humor was not one of his God-given gifts.

"Thanks a bunch," Nick said with a nod and left.

Ricardo watched the Inspector walk down the hall and into the stairwell. He looked at his watch and saw that it was six-thirty, time to take his medication. He took a pill out of his pocket and washed it down with a sip from the water fountain in the hallway. Since it was still relatively early, he thought that maybe he should try to sift through at least one of the drawers in Eduardo's filing cabinet. He headed back down the hallway toward the door left unlocked by the Inspector.

Entering the room his eyes immediately fell onto the space where the computer should have been. Damn it. He knew that the professor undoubtedly had dozens of equation solvers already programmed that he would have loved to have gotten his hands on.

He set his briefcase on the floor next to the desk and opened the top drawer of the filing cabinet. At first glance he noticed that the tabs were all labeled with course numbers. Leafing through the first one, he noticed that they were the Professor's notes on the subjects, but the pages weren't labeled for each day as was his own method.

Most professors had standard notes for each day of lessons which would be used over and over again, but what he held in his hands were simply general notes and suggestions for learning the material.

Amazing, Eduardo had taught from memory. The material must have been child's play for him.

He scanned a few pages and realized that he had some very useful information before him and that he should save the notes for his own teaching.

He looked over his shoulder to the bookshelf and was reminded of the binders, many of which were also labeled with course numbers. He located the one with the same number that he had just leafed through from the filing cabinet. The binder contained sample problems, which had probably been used for homework assignments or for exams. These would come in handy, too.

Ricardo continued flipping through the binder, looking for the solutions, then realized there weren't any. It figures. The man had solved everything spontaneously. He slid the binder into its place on the bookshelf and headed back to the filing cabinet.

He then pulled up a chair and opened the middle file drawer. This one was only about two-thirds full and was neatly organized by conference title. Each folder contained single sheets, which appeared to have been torn out of the conference manuals, each describing a research topic that was apparently presented during the conference. They must have been topics that Eduardo had found interesting and probably wanted to pursue, or perhaps retain for later reference if he needed to contact a fellow colleague for insight on a particular topic.

He slid the drawer shut and opened the bottom one, at

which point a smile formed on his lips. It was the Professor's own research. This would be interesting.

Maybe it was a good idea after all that he should clean out the office. Now he would have access to all of the Professor's research notes and work-in-progress, or more accurately, work-out-of-progress, considering the Professor's recent fate.

Ricardo snickered to himself.

The first several file folders were organized by research students' names. Eduardo had likely begun the research on his own and then solicited students to carry out the bulk of the work, guiding them along the way, then together they would publish the results.

He found the file on Ramon. He would certainly need this to become more familiar with the topic, so he pulled the folder from the drawer, but then remembered the Inspector's warning to leave everything in place. He wasn't concerned. He would simply slip it into his briefcase and review it at his leisure.

The Inspector would never miss it.

The next folder contained the work from another graduate student researcher to whom he had not yet spoken. The student had been away during the week, and had probably not even heard of his advisor's death. He leafed through the file to become acquainted with the topic and noticed that each page of notes was labeled in the upper right-hand corner with the student's name, underneath which were the Professors initials.

The man was certainly organized.

What the hell, he thought, as he decided to simply pull the files of all six current graduate students from the drawer and slip them into his briefcase. He would read through them at home that evening and return them the next morning... if he felt like it.

He couldn't be expected to hang around the office all night. Besides, staying at work late on a Friday night would certainly not help his reputation. It made him appear as though he had no life outside the office, even if it were essentially the truth. And now that he was to replace Professor Rodriguez, he should really try to improve his image among the students and staff.

They were in such awe of the late professor, every word the man said was taken as gospel. All of Eduardo's classes were filled to the maximum enrollment possible and they were always held in the largest lecture halls. Ricardo knew he would never fill Eduardo's shoes, but he would try his hardest to win over the students and to gain some of the respect that he had been lacking from his colleagues by always having been considered second best.

He brought his thoughts back into focus and began leafing through the files in the back of the drawer. The Professor's latest research files which hadn't yet been associated with a student. Ricardo felt as though he had just stumbled across a buried treasure as he flipped through the hand-written pages of complex equations. He couldn't wait to dive into them, but with a huge dose of self-control he put them back into their place. He decided that he would look at them another time, when his mind was fresh, and he could really enjoy them.

He slid the filing cabinet drawer shut, then grabbed his over-stuffed briefcase and shuffled through his pocket for the office key. He turned off the lights and locked the door.

Almost expecting the Inspector to be lurking around the corner waiting for him, Ricardo placed the briefcase on the floor and walked into the stairwell to be sure that he was alone. When he realized that there was no one there, he cursed himself for being over-cautious. Besides, he had done nothing wrong! And he was only doing his job. He shook his head at his own foolish actions and left the building.

Upon arriving home, Ricardo noticed the red light blinking on his answering machine. It was a message from Senhora Gonzales, his housekeeper. She had called to say that she wasn't feeling well and couldn't come to work that day. Therefore, he realized, he would have to cook for himself that night.

He had hardly recognized her voice on the machine. In the twelve years that she had been working for him, they had exchanged very few words. They were seldom in the house at the same time, but when they were, she had always busied herself with her work and seemed to avoid any type of conversation with Ricardo. That suited him just fine, as he couldn't ever find much to say to her other than commenting on the weather. Once in a while he would leave a note for her about food preferences or to let her know if he would not be at home for dinner and she shouldn't leave anything, but other than that, they lived in separate worlds.

After taking off his coat and tie, Ricardo fixed a scotch and soda and walked out onto the back terrace. The onset of darkness was approaching and he could hear the frogs chirping in the pond behind his house. It was a small but peaceful backyard. He sat down in one of the two patio chairs, let out a sigh, and took a sip from his glass, immediately feeling relaxation wash over him.

He began to think about his postulate, something that he hadn't given much time to over the past week due to Eduardo's death. Perhaps a break from it had been a good thing. It had consumed his thoughts for the last several months, and now he might be able to sit down with a clear head and make some progress on it.

He knew that he'd be spending the night going through the graduate students' files, and then tomorrow he'd head back to the university to replace what he'd taken and continue sorting through Eduardo's office. But on Sunday he would focus on his own work.

It was now more important than ever that he begin publishing some research and fortifying his reputation as a top mathematician. Then maybe students would come to the university specifically to study with him, instead of Eduardo.

Ricardo returned to the kitchen and opened the refrigerator to see if there was anything he could find to eat. There were not many options as he almost never had to cook for himself, nor shop for that matter. Senhora Gonzales normally did her cooking at home and dropped off his serving when she came by to take care of the other household chores.

He found a dish left over from the night before with some rice and beans, which would have to be enough to carry him over until the morning when he could go out for breakfast. Nibbling on the cold leftovers while standing at the kitchen counter and bored with his simple meal, he decided to take a look at the students' research files while he ate.

The first file that he pulled out of the briefcase was for the one student with whom he had not yet spoken. The topic was prime number theory. It was a subject with which he had

had little experience other than knowing the basics. How was he to strengthen his reputation when his students knew more than he did about their research topics?

He realized that he would have to spend considerable time in the near future learning all that he could about these new topics. As much as he enjoyed his current field of expertise, he couldn't limit himself to it. He would have to expand his knowledge base.

Maybe it was a blessing in disguise that Eduardo passed away so suddenly. It was as if he was unexpectedly handed the opportunity to pursue other fields of study. Maybe he would even find new topics to conduct his own future research.

He tossed the file folder onto the floor and reached over to pull the other five from his bag. As he did so with one hand on the files and the other on his fork, the folders slipped away and dispersed across the kitchen floor.

He cursed under his breath.

But, in consolation he thought, at least the good professor had been extremely organized by labeling each sheet with the students' names and page numbers.

He took another few bites of beans and then laid the empty plate in the sink. He began to collect the papers and return them to their proper places. Most of them had retained their order but a few had gotten free and slid across the room.

Leaning over to grab the loose sheets, Ricardo noticed a few pages, paper-clipped together, which had no name in the upper right-hand corner. He took them in his hand and sat down on the floor to try to determine if he could recognize the subject and file them in the appropriate place.

A first glance at the contents gave him a feeling of déjà vu.

Then he realized why. What he held in his hands looked similar to his own research. It contained the workings of a growth model. He couldn't believe his eyes. Did Eduardo have a graduate student working in his own field of expertise? That would have been unheard of since the students always seek out the professor who is most knowledgeable in the respective field of research.

Trying to recall the six student's research topics, Ricardo realized that none of them had any use for this type of analysis.

Could this possibly have been some of Eduardo's private work? If so, why had Eduardo not discussed it with him? Ricardo had often gone to his colleagues for advice in their particular fields of expertise. And why hadn't Eduardo ever mentioned an interest in his own area of specialization?

Ricardo couldn't make any sense of the situation, but then realized that, without Eduardo around to clarify, he would probably never get the answers he sought.

But what was even more intriguing at the moment was finding out what Professor Eduardo Rodriguez had been trying to determine with these equations and pages of notes.

He took the sheets into his study, sat down at his desk and turned on the lamp. Ricardo noticed that his hands were shaking as he laid the papers down before him. Why was he so nervous?

He took a deep breath and then began to read. The first couple of pages looked as though they were the workings of some complex variables. There were several single, double and triple integrals, solved over a multitude of constants and variables. It looked as though the whole Greek alphabet had been used in definition of the inputs.

The last page was apparently a summation of the variables on the previous pages—the equation in it's whole—in the form of a linear regression model. But whatever it was supposed to represent, Ricardo didn't yet know.

Sitting back for a moment, he took a deep breath and tried to focus his thoughts on his mathematical abilities, rather than trying to solve the mystery of why Eduardo had been dabbling in a field in which he had openly shown no interest.

He was so excited and curious to find out what Eduardo had been attempting to solve that he had to force himself to think rationally and, most of all, calm himself down.

The logical steps to understand what lay before him were to get a feel for what the equation was trying to determine and then analyze each individual element.

He flipped again to the last page and began studying. There was some scattered commentary in the columns which looked to define a few of the constants and variables. He began taking notes on a separate sheet of paper, listing each of the symbols and defining them as he deciphered the scribbling.

Lambda apparently stood for latitude, while Kappa was longitude and Alpha was altitude, or more likely, height above sea level.

With the definition of these variables, Ricardo thought perhaps that Eduardo had been trying to create more accurate population growth models for different locations on the earth. These were used to predict population numbers in years, or even decades to come, which could be balanced against natural resource availability and environmental planning.

As he flipped back through the preceding pages, looking for the variables in question, he partially confirmed his suspicions by locating their related functions and recognizing their impact on the summation equation.

But looking again through the notes, he realized that he had only touched the tip of the iceberg in putting the whole puzzle together. There must be dozens of variables before him that he didn't recognize! How frustrating!

What was this that Eduardo had been working on?!?

He released the pages from their hold in the paper clip and laid them out adjacent to one another on the desktop. There were six sheets in all.

Once again focusing his attention on the last page he tried to decipher more of the variables. He noticed that about half of the equation segments were negative, illustrating decay; therefore, it wasn't strictly a growth model.

Natural growth and natural decay in the same equation?

That was certainly interesting but didn't make much sense. Normally, one would attempt to determine one or the other exclusively, but not both simultaneously!

His own studies had mostly focused on population growth functions which had many factors including agricultural and medical advances.

Whereas, the most well-known example of decay analysis was using one of the isotopes of the element Carbon. Carbon 14 is radioactive and decays at a rate proportional to the amount present. It is currently the most popular method used to date ancient artifacts. But that didn't appear to be the case here.

So, how do these opposing elements interact?

Focusing on the notes in the margins he could make out a few more variables. It appeared as if Rho stood for Race. That would support his previous conclusion of optimizing growth rates in different regions of the world, but wouldn't that then be redundant? Precise world locations would, as a general rule, imply which race inhabits the area. Unless he is trying to include multiple factors for all varying races in the region? Could that be possible?

A hand-sketched arrow pointing to the variable Omega had the word 'height' at its trailing end.

It was nearly unimaginable, but could it be referring to the height of human beings?

The variable did not appear to be integrated over an entire population, therefore, it was referring to a single entity. A single human being. The growth and decay of one person.

My God!

Was it actually possible that what he held before him was an estimate of individual human growth and decline?

The lifespan of a single individual?

No, that's impossible, he thought. There must be something missing!

Ricardo quickly gathered the papers together, stuffed them into his briefcase and headed for his car.

Chapter Eleven

The precinct was unusually quiet for a Friday night. As Nick walked towards Officer Valdez' office he remembered that it was still early, only seven o'clock. It must be the calm before the storm.

Since he got his promotion, he no longer had duty on Friday nights. He was only in the office in the evenings or weekends when it was necessary, and today it was merely to drop off the computer. As soon as that was taken care of, he'd be quickly out the door and on his way home for a quiet weekend with the family.

His colleague, Manuela Valdez, sat at her desk typing away at the computer as he entered her office. Completely engrossed in her work she didn't seem to notice him. Nick waited to see if she would detect his presence, but she continued clicking away, intently concentrating on her screens. She then leaned over to peer more closely at the monitor to her right, and then back to the terminal in front of her. Nick cleared his throat and she immediately turned around.

Frowning at the sight of him, she looked at her watch. "It's ten after seven on a Friday night. What the hell are you still doing here?" She then continued with her typing.

Nick placed the computer on her desk in the only available space amongst the four other terminals and several partially disassembled hard-drives.

Manuela was a wizard when it came to computers. Not only was she a software expert, but she was an ace with hardware too; she could disassemble and re-build any computer model, and make them do almost anything but serve you a cold drink. Her life was her work and she paid little attention to much else, including her appearance. She was slightly overweight and seemed to wear clothes that were always a bit undersized. At thirty-eight years old, with mousy brown hair cut in a short bob and thick-rimmed glasses, she was single and would likely remain that way.

"I came to see if you were free this evening." Nick liked to tease her and she had a dry sense of humor which enabled her to hold her own ground among her mostly male co-workers.

Without even looking up from her computer she rattled off an answer. "What did you have in mind? Not another mud bath, I hope. I'm still turning down offers from the other five guys we shared the tub with."

"Rough life you lead," Nick said with a slight smile. Then he turned the conversation to business. "Manuela, I need you to work your magic on this box."

She looked up from her work and gave him her full attention. "What's up?"

"Well, did you hear about the mathematician who died last week at the university?"

"Yes, I heard about it."

Nick sensed immediately that the topic upset her, as she turned her head to one side in an attempt to hide her emotions from him.

But before he had a chance to say anything more, she spoke up herself.

Pointing to the computer she asked, "Is that, or *was* that his?"

"It sure is. We have reason to believe that, well..." Nick hesitated, trying to find the right words, though he had no logical explanation himself. "Well, we're not quite sure what happened yet, but there is a file on here entitled *Upon_My_Death* that needs a password to open."

"Pretty spooky." She said faking a frightened look. "Where's it located?"

Nick was relieved. She seemed to be back to her old self with that comment. So he sat down and started talking. "I found the file in the Documents folder. I'm not really sure where it's located on the hard drive, or maybe it was on a disk or stick or whatever those things are called."

"No problem, I'll find it." She wrote Nick's name in large block letters on a Post-It note and slapped it on the box. "Anything else?"

"Well, just see what you can find. There are hundreds of files, most of which are not in Word, which is about the only thing I can make sense of. Besides, the guy died under relatively suspicious circumstances, so if you come across anything interesting, just let me know."

"*Interesting*? What would you find *interesting* to a mathematician?"

"Good question. I'm not really sure, but I have an uneasy feeling that something he may have been working on is linked to his death."

Again that look. But she continued, "Uh oh, when you

have an 'uneasy feeling' then I know there's work to be done." She took a gulp of coffee from her mug and sloshed it around her mouth before swallowing. "Can this wait till Monday or do you want me to call you at home over the weekend if I find anything? I can disguise my voice as a man's if your wife answers so she doesn't suspect our clandestine affair."

"Good idea. We can never be too careful." Nick played along, although the thought of an affair with Manuela really wasn't very funny, even to joke about.

Then, as an afterthought, he added, "You'll be working over the weekend? Don't you ever take a day off?"

"Very funny. Should I call you or not? Otherwise I've got lots of other work..." She gestured to the stack of equipment on her desk.

"Yes, call me." Nick then got up to go. "Oh, and since you are so busy, I guess we can just postpone the mud bath 'til next weekend. But the guys will be disappointed."

Manuela was already back to her computer work as he was walking out the door but managed a half-hearted reply. "Sure. I'll mark it on my calendar."

Sonja greeted him at the door as he arrived home. She gave him a quick kiss as he stepped inside and then Nick was practically overrun by his two sons spilling out ideas for the weekend ahead.

"I want to go fishing!" Shouted his older son, Alvar, who had caught the bass the previous weekend.

"We just went fishing last weekend. I want to go get a dog!" retorted his seven-year-old, Matt.

Nick was surprised by this turn of events. "A dog? Where

did you get that idea?" Looking to his wife for support, she only shrugged her shoulders.

Matt explained his desire for a new pet. "Rico got a dog last weekend and I was at his house this afternoon playing with it."

Nick and Sonja looked at each other in understanding.

Matt continued with his negotiating tactics. "It's really fun to play with and you only have to feed it twice a day, which I could do myself, before school and at night. Plus, it's good to have a watch dog!"

"What kind of dog does Rico have?" Sonja asked.

"I don't know, it's brown and well, has some other colors mixed in. It's really cute!"

Nick grabbed his young son around the waist and carried him upside-down into the living room. All the while, the boy was giggling with delight.

After setting him back on two feet, Nick knelt down to his son's level, hoping to diffuse the dog situation as quickly as possible. "Son, a dog is a lot of work, and since no one is home during the day, that would mean it would be all alone. I really don't think that is fair for an animal, do you?"

"No, but...." Matt thought for a moment.

"But what?"

"But it wouldn't mind, not if it had me to play with in the afternoons!"

"I'll tell you what. Why don't you plan to play one afternoon a week at Rico's, if it's OK with his parents. That way you can have fun playing with a dog, but we don't have to feel sorry for one sitting alone at our house all day."

"Well... but... if I still want a dog of my own, can I ask for one for Christmas?"

"You can ask for anything you want," Nick said to the surprised eyes of his son, "but that doesn't mean you get everything you ask for."

Matt shook his head in disappointment. "Figures," he said as he stomped off to his bedroom.

"Good save." Sonja applauded her husband. "For a minute there I thought we were going to have a new furry family member."

Nick smiled. "Dogs are more work than kids and you can't just pick up and go away for the weekend. That's exactly what we don't need right now."

He followed his wife into the kitchen where he got a whiff of something delicious coming from the stove. "What are we having for dinner? It smells great!"

"The kids already ate some pasta." She took him in her arms and gave him another kiss. "I made a special dish for just you and me. We can eat alone while the kids watch a movie."

"Sounds irresistible." Nick tried to snatch another kiss but his wife wiggled away to get back to her cooking. He watched her move through the kitchen. She was lovely. Even after two kids her figure was trim. She was quite tall and had soft brown hair, cut short in a sporty look, which suited not only her physique but her personality perfectly. He leaned against the kitchen counter next to her. "Is there anything that I can do to help?"

Sonja gave him a smile and a sarcastic laugh, which likely meant that his help in the kitchen would probably lead to inedible food. "Why don't you tell me about your day."

His head dropped in mock anguish and he let out a loud sigh. "Can't I ever get away from it?"

She knew her husband was exaggerating but she was actually interested in his work. "It can't be all that bad. There has to be a bright spot somewhere."

Nick thought for a moment then answered her truthfully. "Actually, I am working on an interesting case."

Looking at him with surprise since he never uses the word 'interesting' in connection with his work, Sonja responded, "Great. Then let's save it for over dinner. It'll be ready in about five minutes. Maybe you could get the boys started with the movie, while I get it on the table."

"Deal! Back in five ... no more!"

A few minutes later, Nick opened a bottle of wine and poured out two glasses. He then began telling Sonja all about the death of Eduardo Rodriguez, the odd Professor Montagne, the grieving Rodriguez family and the less-than-polite priest at the family church. She giggled through most of the story as Nick had a way of dramatizing even the most mundane events.

"So, who did it?" she finally asked.

He looked up in surprise. "Who did it? Almighty God, I guess. I mean, you know as much as I do. Do you really think that it was anything other than death by natural causes?"

"Sounds fishy to me," she replied, wriggling her nose as if she were sniffing a stinky fish.

He played along with her word game. "Is that a subtle hint that you want to go fishing again this weekend?"

She laughed. "Actually, last weekend was fun. I wouldn't

mind doing it again sometime, but what would we do with the dog?"

Nick balled up his napkin and tossed it playfully at his wife's grinning face. They both broke out in laughter.

Chapter Twelve

Ricardo was out of breath from his race up the stairs when he unlocked the door to Eduardo's office. He tossed his things down on the desk chair but didn't take the time to settle his breathing before tearing open the filing cabinet drawer which held the personal files that he'd seen before. He pulled out a stack of folders and quickly began thumbing through them. After glancing over each one he realized that nothing resembled what he was looking for, so he sat back, took a deep breath and slowly began reexamining each page to be sure that he hadn't missed anything.

He thought for sure there must be something more on Eduardo's mysterious work that he'd found in those six sheets of hand-written notes. There must be some clarification or a global key to the variables. He must have attempted to apply the equation at some point, and thus the magnitude of the numbers could give him a better feel for what was trying to be determined. But where are those numbers?

He then turned his attention to the remainder of the filing cabinet. When nothing turned up there, he began leafing through every book, every file, and every binder as well as each and every scrap of paper that existed in the office.

After countless hours of searching, he realized there was

absolutely nothing in the office that could provide further insight into the equation that he had come across earlier at his home.

He felt exhausted.

He sat back in the swivel desk chair and turned to the window to see the sun rising. It was 6:45 a.m. Had he spent the entire night in the office on a wild goose chase?!?

"Where was the rest of that damn equation and the variable definitions?!?" Ricardo wailed out loud.

They must be on the computer the Inspector took!

How the hell was he going to get his hands on that?

But then another thought came to him. Maybe, just maybe, the Professor did some of his research at home. What if the files were stored on his home computer?

The sudden thought that an over-eager student may soon stumble into the building on a Saturday morning got him to his feet. He didn't want to be found in the position where it was quite obvious that he had spent the night in the office. Gathering his jacket, he searched for the office key only to find it dangling from the keyhole where he had left it in his hurried attempt to get in.

On Saturday morning Nick lay on his couch, deeply engrossed in the sports section of the paper when his phone rang. His hip cracked as he rolled over to reach for it.

"Nick, it's Manuela." He heard the voice whisper on the line.

"Manuela who?" Nick spoke in all seriousness even though he recognized her voice despite the whisper.

"Valdez. From the precinct." She continued in a whisper.

"Manuela. Why the hell are you whispering?"

"Well, I didn't want to disturb you at home."

What?!? But Nick thought it better not to ask. "Manuela. You're not disturbing me. What's up? Did you find something on that computer?"

"Well, yes and no."

"Go on."

"The *death.doc* is coded. I got it opened but can't read it."

"Why not?"

"It's encrypted."

"Damn it."

"Yeah, well, don't get too upset yet. We can probably get it de-coded. I've got a nephew who is a genius, and that's saying a lot coming from me you know."

Nick chuckled. "Takes one to know one I guess."

"Yeah, something like that. Anyway, I'd like to get him to take a look at it, but I wanted to get your OK about having a third party handling the computer."

"Sure, whatever you need to do. The way it stands now, no one can look at the files, so we need to do something." Nick sat upright on the couch. "Did you find anything else?"

"Actually, that's the main reason for my call. There are a couple of files that had been deleted just before the professor checked out last week."

"'Checked out?' You're a cold woman, Manuela."

"Only at the office, hon."

Nick stifled a cough as an excuse to not have to reply to her remark with words. "But what's so interesting about some deleted files? The guy tried to get rid of just about everything. He seemed to have wanted to leave a clean slate."

"Just listen for a sec. They appear to be some techno-equation stuff. It's not in my field so I don't understand it." She paused when she heard Nick sigh. "But what really caught my attention was not so much the content but the titles of the files."

Nick was intrigued. "Which were?"

"One was called 'Lifespan' and the other 'My Death'."

Through a grin Nick added, "Please tell me that your nephew is a genius at 'techno-equation' stuff too."

Manuela snorted. "To him, this stuff is a walk in the park."

It was finally Monday and it had taken all of Ricardo's self-control to keep himself from going to Eduardo's house on the weekend when he was sure to find family and friends hanging about.

Ricardo rang the doorbell at the Rodriguez residence just before noon. Rosalinda answered after the hurried second ring and appeared well-kept, as always. She smiled warmly when she saw her husband's colleague and immediately invited him in.

"Have you eaten yet, Ricardo?" Despite their infrequent social contacts, Rosalinda was always kind and gracious to him, and they were on a first-name basis. "I've just been preparing tostadas. If you have a few minutes, I can have them on the table and we can chat over lunch."

A slight guilty twinge rang through his body and, despite his eagerness to get down to work, he accepted her offer and looked forward to a pleasant meal, especially after eating next to nothing all weekend.

Rosalinda bustled about and, instead of using the large oak table, she set the small table in the kitchen with two place settings. "It's so nice that you're here. Now I don't have to eat alone. Please sit." She pulled out a chair for him.

"So, how are you doing?" he asked with all sincerity once they were seated. She looked just the same as always, she was not beautiful, but she had an honest, wholesome face.

Her eyes appeared wide and dark, as if bottomless pools. "Honestly, I feel empty," she said. "I feel very sad and I miss him. It's that simple."

Her candidness surprised him. He was not used to talking about emotions. "It will take some time," was all he could summon for comfort.

"I know, that's what everyone says." She sighed and took a bite of food. After she was finished chewing, she added with a smile, "At least I have Maria near me. I don't know what I would do if she lived far away from me like Miguel. I need one of them here."

"Speaking of your children, how are they handling it?"

"As well as can be expected, I suppose. We're in mourning," she sighed, "but fortunately for both of them, they have jobs to go back to, which help to keep their minds off of things."

"Have they been working this past week?" Ricardo was a little surprised to hear that the children had so quickly resumed their normal lives after the untimely death of their father.

"Maria was with me all of last week and finally returned to work today, same with Miguel, he left just yesterday." After a deep sigh and a genuine smile she added, "Not to mention that he has his own family and his wife is pregnant, they need him more than I do." Taking a sip a water, she shooed a fly from the table edge. "Of course, I would have liked for him to stay here for a bit, help me sort things out, but he loves his work and I knew it would be the best thing for him to get back to it. We all have our own ways of healing."

He felt obligated to ask what her way of healing was, but he was much too eager to change the subject and get down to his reason for being there. "Rosalinda, I've taken over much of Eduardo's responsibilities at the University."

"Of course, that only makes sense," she replied.

Trying to determine just how he could delicately approach the topic, he bought a bit of time by eating another bite and complementing her on her cooking.

As casually as possible, he continued, "The police were in Eduardo's office last week and—"

"Oh, no. Oh, no. I'm sorry Ricardo. I'm not sure why they are causing such a stir." She was visibly upset.

He tried to calm her immediately. "It's alright, Rosalinda, there was only one Inspector there, and it was after school hours, so none of the students were affected."

It's now or never, he thought. "It's just that they, the police that is, took his computer," now comes the punch-line, "and there were some files on it which are essential to the research of some of his graduate students."

"Oh! Eduardo would be so upset if he knew that his students' research was disturbed. Is there anything I can do to help? Should I talk to the Inspector?"

"Well, I'm not sure if that would do any good considering that it was not personal property. The computer did not belong to Eduardo, rather to the University."

"Oh." She was out of ideas.

Ricardo would have to spell it out.

"But maybe he had stored some of the files on his home computer," Ricardo said, "as a backup. And I was wondering… would it be possible for me to take a quick look at it? If he has some of the students' work stored here, it would be a big help for me to read up on it."

She seemed to immediately perk up at the suggestion. "Oh, of course, Ricardo, no trouble at all. He did spend quite a bit of time on his computer, you know. Oh, I'm sure he has those files here, he was always so careful and orderly." She stood up and began bustling from the room. "Right this way, Ricardo."

But there was no need for cajoling as he was already right on her heels.

Chapter Thirteen

The thin red digits of the alarm clock read 3:45 a.m. as Ron slipped out of bed in the dark. He stood under the shower head with his eyes closed, still half asleep, enjoying the heat of the steamy water. As he opened his eyes, he saw Caroline standing in the middle of the bathroom watching him. He quickly shut off the water and opened the glass doors.

"Is everything all right?"

"Of course everything is all right. I just wanted to spend a few minutes with you before you left this morning." She shook her head at his over-reaction. "Besides, I brought you some coffee."

After toweling off, he took the cup of steaming black liquid and enjoyed the first sip. "Ahhh..." he said, then began getting dressed.

"So, Joe is picking you up, right?"

"Yes, at 4:30."

"And what time is your flight? You're flying out of Logan Airport, right? Not Greene?"

"No, well, yes. Actually we're taking a flight from Providence to Boston then another plane to Rio. The Providence airport doesn't have direct flights to Rio, and then we change planes there for Brasilia."

"What time?" Caroline asked impatiently.

"What time do we change planes?" Ron asked, then realized he hadn't answered her question. "Oh, the flight this morning is at 6:30. Don't worry honey, we have plenty of time to get there and check-in and probably even enough time to have a five-course meal before taking off." He smiled, but his wife didn't follow suit and he knew that she wasn't in the mood for jokes. "Should I stay here?" he asked.

"Oh, God no. Of course not. I'll be fine for five days. I'll just miss you, that's all."

"Honey, are you sure that you are OK. I mean, this conference is important to me, but when it comes to my family it means absolutely nothing. I'd stay here in a heartbeat if you want me to."

"No, no." Caroline said almost in a whisper with her eyes turned down like a shy girl. "I'll be fine as soon as you are gone. I just hate good-byes, that's all."

"This isn't a good bye, it's just a see-ya-later."

Caroline let out a little laugh and Ron knew she would be fine. "I'll call you as soon as I arrive, and you have all my hotel information so, if you need to reach me, you can, anytime."

"Oh, I'm jealous! I wish that I were going with you." She began pacing the tile floor. "We really should travel more. I've never even been outside of the US, well, except for Canada, but there isn't such a big difference between here and there. I think we need to see more of the world, different cultures, don't you think?"

Ron knew where this was coming from. She was probably thinking that if her time left on the planet was limited, she needed to see and do as much as she could right now.

At the moment, all he could do was agree with her. There was no use in getting into a deep conversation at 4 a.m. "Alright, why don't you decide where you want to go, and when I get back we'll start planning the trip."

"Really? Oh, that would be great." She wrapped her arms around Ron's steaming body and held tightly, then began kissing his neck.

"Caroline, you had better stop that or I won't be ready when Joe gets here."

She pulled away from him, smiling. "You finish getting your stuff together and I'll put breakfast on the table, just something quick."

She was gone before he had a chance to object. He'd have plenty of time at Logan to get breakfast, but if it pleased her to send him off with something in his stomach, then eating twice was a small price to pay.

The flight from Providence to Boston was over in what seemed like a few minutes and both Ron and Joe wondered why they didn't just drive the extra hour to Boston.

When they got to the gate for their next flight, they had the pleasant surprise of being bumped up to business class because of an overbooking.

Business class service offered him yet another full breakfast. How many was that so far? After they cleared his tray, he reclined the seat and took a short nap. Awaking refreshed, he and Joe talked about their plans to investigate the city layout and buildings of Brasilia in addition to attending as many of the conference presentations as possible.

Other than being long, the flight to Rio was uneventful, and after a short layover there, they were finally on their last leg to their destination. The skies were clear and Ron had a great view of the countryside from his window seat, where he sat perched like an anxious child. At 7 p.m. they began their descent into Brasilia.

Flying in from the north, the landing pattern had them first circle around the city by crossing over it and coming back from the east. This gave Ron an opportunity to see the entire city from the air. He recognized what he thought was the Parque Nacional de Brasilia, one of the only places in the city with green spaces, lush gardens and natural swimming pools.

He was surprised to see how orderly it looked for such a highly populated place. In school, he had learned that the city was actually planned to be laid out in the form of a bird in flight. And from above, he could vaguely make out the form. The head of the 'bird' was pointing towards the giant man-made lake, Lago do Paranoá, with its wings comprised of row after row of apartment buildings. The 'tail-section' encompassed three sprawling parks, while the body held the expanse of what he had read were mostly government buildings.

After landing, a portable set of steps was rolled up to the plane and, as Ron stepped out, he was immediately enveloped in the warm air. It felt great breathing unfiltered freshness after being in the air-conditioned plane for the entire day. He took a deep breath and descended the stairs.

They passed through customs without incident and picked up their luggage. During the cab ride to their hotel, which was also the location of the conference, they got their first real taste of the city up close.

The streets were bustling with cars, taxis and small transport trucks, but the buildings looked barren, surreal and futuristic with extravagant arches and sloping braces. They looked aesthetic but cold. Shops and cafes were few and far between, as were pedestrians. Although there appeared to have been an attempt to plant some trees amongst the buildings, there was no grass, only reddish crusty earth, baked by the sun to hardness that matched the concrete sidewalks running through them.

The stories that Ron had heard about Brasilia were not exaggerated, the city really did have a sterile feel to it. At least that was his first impression. It didn't resemble any of the cities that he'd seen in the US in terms of architectural styles. He felt like he was seeing a whole new world. Actually, the perfect place to work and learn for a few days.

After checking into their rooms, Ron fell into a comfortable chair and dialed home. "Hi honey, it's me!" he said on hearing his wife's voice.

"Hey, I didn't expect to hear from you so soon. How was your flight? Did you get enough to eat?"

He smiled at her concern. "Yes, I ate plenty on the plane and the flight was fine. We were bumped up to business class so I even got to recline my seat and sleep a little."

"Wow, aren't you important! Business class, no less!" she responded, teasing him. "So, what's the city like?"

"Well, we just arrived, so we haven't seen much except for the cab ride to the hotel and, well, it's pretty clear 'I'm not in Kansas anymore'."

Caroline laughed. "Be sure to take lots of photos."

"Okay, I will. So, how are you feeling?"

"I'm fine, a little tired, but nothing new." She sighed, then added, "Please try not to think about this while you're there. Just enjoy yourself. I think this is a good opportunity for you to clear your mind and relax."

"I'll try." Ron could hear Sophie in the background saying something to her mother.

"Ron, Sophie wants to talk to you so I'll say goodbye. I'm glad you called and it's really good to hear your voice."

"Yours too. I'll give you a call tomorrow if I get a chance. I love you."

"I love you too. Okay, here's Sophie!"

Ron could hear some shuffling noises as his daughter was trying to get the phone positioned properly.

"Hi Daddy!"

"Hi sweetheart. How are you doing?"

"Good! I went with Mommy and grandma and grandpa to McDonald's for dinner!"

"Wow, I wish I could have been there too!"

"Don't they have McDonald's in Africa?"

"Africa? Sophie, I'm in South America, in Brazil."

"Oh, don't they have McDonald's there?"

"Well, I'm sure they do. I'll have to look for one and then I'll think of you!"

"Okay, maybe you can get a rhinocer-saurus burger!" She giggled.

Her laughter brought a smile to his face. "Maybe. I'll have to check out the menu. Well, Sophie I have to go now, but you take care of your mommy this week and I'll talk to you real soon, okay?"

"Alright, daddy! I love you!"

"I love you too, Sophie. Bye."

"Bye, daddy!"

He ended the call home, then picked up the hotel phone and dialed a three-digit extension.

"Hey, Joe. What's the plan?"

"The plan is food."

"Yeah, I'm pretty hungry too, amazingly, since all we did was sit around and eat all day, but I'm also exhausted."

"Same here. Want to grab something quick in the hotel restaurant and then call it a night?"

"Sure. Actually, I'd like to take a quick shower first." He paused to look at his watch, it was five minutes before nine. "How 'bout if I meet you in the lobby in say... twenty minutes? At about 9:15?"

"Sounds good."

Joe was waiting in the lobby, reading what was posted on an announcement board, when Ron exited the elevator. All of the companies attending the conference were listed on the board, along with activities and events sponsored by both the conference sponsors and the hotel. Ron grabbed a leaflet from an adjacent table, which contained all of the same information, and then they headed to the restaurant, passing by a noisy, smoke-filled bar on their way.

The hotel restaurant was apparently designed to match the style of the city with sweeping columns and exaggerated curves. They were seated at a table over which hung a photo of Lucio Costa and Oscar Niemeyer, the architectural founding fathers of Brasilia. Both Joe and Ron recognized the two men from college textbooks without having to read the bronzed plaque fixed below. All of the walls were covered with large photos of the city's most magnificent buildings.

"Looks like we can get a city tour just by staying in this room," Joe said, admiring the décor.

They each ordered one of the local beers and opened up the menu.

"I feel like I'm in the twilight zone," Joe remarked without looking up from his menu.

"Yeah, I know what you mean. This place doesn't give you that warm homey feeling, does it?"

Both let out a laugh, but Ron's own comment reminded him of home and how that 'warm, homey feeling' was something that had been missing since they'd gotten Caroline's diagnosis.

Almost reading his thoughts, Joe asked, "How's Caroline?"

Ron put his menu aside, having decided to order a steak. Although the question was simple enough, it weighed him down. Perhaps amplified by his fatigue, he sighed deeply.

Surprised by the reaction, Joe added, "Whoa, buddy, I didn't mean to get you upset."

"No, sorry, it's not you. I'm just tired. Caroline is doing fine, or as well as can be expected, I guess. It's just tough on both of us. There are so many unanswered questions. It's like we're living from day to day."

Ron looked up and saw the blank look on Joe's face, who clearly didn't know what to say in reply. Joe was younger, in his late twenties, single and had somewhat of a reputation for being a ladies man at the office. They had worked together for three years and Ron knew that Joe came from a large Italian family with good values. So despite their obvious differences, they were similar in many ways and got along well, and when he got the news about Caroline, he had shared it with his friend. "But hey, let's not talk about that right now. What about you? How was your date with 'Miss Narragansett Bay' on Saturday?"

Joe blushed slightly and took a sip of beer while trying to hide his embarrassment. "Her name is Suzie. It was fine."

Ron waited for more, then realizing that was the sum of the response, he pushed on. "Fine? That's it? What kind of a date is fine? Give me the goods!"

"There's no goods. We had a nice time, she's a great girl, not to mention gorgeous."

"And? When's the next date?"

"I'm supposed to see her again on Saturday night."

"Well, well." Ron sat smirking at his friend.

Joe was fidgeting with his napkin now. "What?"

"You're in love!"

Joe waved his hand in Ron's direction in a futile attempt to signal his lack of feeling about the matter, but Ron continued. "Wow, this is gonna make headlines back in Rhode Island. The state's most eligible bachelor is off the market!"

"Nah... Just because I'm taking a girl out on a second date doesn't mean I'm *in love*," Joe replied, drawing out the words. "Actually, we had a good time. That's all."

Just then, the waitress approached their table, ready to take their orders. Surprisingly she spoke excellent English, with only a slight accent. She said most of the customers were tourists, so she had to speak English to get the job. After a brief exchange with her about what she recommended for food, sightseeing, nightlife, etc., both Joe and Ron ordered steaks.

"So, where were we? Oh, yes, your Miss Right."

"Yeah, well, seeing that you are the only one sitting at this table who is married, I guess that means the subject is on Caroline." Joe paused a moment to make sure Ron was on track with him. "Go ahead and talk if you need to get it off your chest. I may not be able to give you much advice, but at least I can listen."

Ron thought for a moment, considering how much good it might do him to talk, but he then reconsidered, "Actually, I promised Caroline that I would clear my mind of it this week and try to relax and enjoy this trip. So, I appreciate your willingness to listen, but I'll have to take a rain check."

"Okay, I'm ready anytime," Joe said, as he was looking around the restaurant. "So, where's that cute waitress. I'm ready for another beer."

At 7:00 the next morning, after already having eaten breakfast, Ron was at reception inquiring as to where his company's display material was being held. He was directed towards the main hall and told that a loading area in the back was distributing the material.

The colossal hall would soon be filled with booths of private companies displaying photos, models, and drawings of their work. Most were manned by employees willing to discuss projects in an attempt to garner name recognition or new business.

He passed through the large conference hall and noticed how it looked just the same as any other conference he had attended stateside. He remembered how exciting it was to participate in his first conference. He had met other architects from around the world and sat at each conference booth examining designs and asking questions. Probably more questions than he should have, taking valuable time from the other firms' representatives and deterring potential new business customers from asking their own questions. But he was young and eager to learn. Now he spends more time explaining his own work, style and ideas to younger architects. He enjoyed the role reversal. Although, he still made a point to briefly view each and every display, stopping longer at those that pique his interest.

He spotted the queue of conference attendees at the receiving area and proceeded to get in line and wait his turn. Then he spotted Joe about ten people ahead of him, so he wandered up to his side.

"Morning, partner."

"Hey! Slept in this morning, huh?"

"Joe, you were probably here five minutes before me and I can almost guarantee that once we collect all our gear you will disappear for an hour or so to eat breakfast. Am I right or am I right?"

"Whoa, slow down, Tiger. Looks like someone didn't get his Corn Flakes this morning?!?"

"I didn't see any Corn Flakes at the breakfast buffet, besides I don't think it's the food that's got me all wound up, their coffee is as strong as hell. Ask them to water it down for you."

"Yup, that'll be the day when Joe Pinochelli waters down his coffee!" He slapped Ron across the back and laughed. "Good one, buddy!"

Ron just rolled his eyes and counted how many were in line in front of them. Just a few more.

In five minutes they were giving their company name to

the distributor and then being handed a floor plan with the location of their display table along with a loaded dolly of material which their company had shipped.

Once they had found their table, Ron began unpacking their stuff and soon realized that he wasn't getting much help from Joe, who was looking at the displays next to them and peering over the partitioned wall behind. Ron knew full well that Joe wanted to check the place out. He liked to wander the halls while people were setting up. He once told Ron that he got a better picture of what the other architects were all about if he could catch them off guard. He would find a group who had already placed some items out for viewing, observe their models and casually ask questions with a cup of coffee in his hand.

"Go ahead, Joe," Ron said with a sigh.

"Thanks, buddy. I've got clean-up detail!"

"No kidding," he responded, but Joe was already out of earshot. The situation was the same every time. By the end of the day, Joe will have met a girl and plan to have dinner with her, so as soon as five o'clock rolls around he will beg Ron to clean up the display stand alone. Ron will make a small scene, then hesitantly accept his fate. Both knew it was a game because Ron preferred to clean things up himself anyway. He was somewhat of a micro-manager and besides, putting away the display only took about half an hour since much of it was left out overnight to be set back up the next morning.

During the first day, they would take turns manning the table, which allowed them each to attend some of the presentations being given in the adjacent conference rooms. Then on the second morning, they presented their own work in a small half-hour forum, which was to be the focus of their trip. After the presentation they would both return to their table for the afternoon, which has always proven to be the busiest time for them. This was mainly because many people who had seen their presentation would come by to ask questions.

At the end of the second day, their official duties were over and on the third and final day of the conference, they would simply have a display with a selection of photos, models, brochures and contact materials. Both would attend presentations and then sit with some of the other presenters and ask more questions and try to acquire as many new ideas as possible to apply to their own work. When they got back home, each would also write a report, a short summary, to share with their colleagues at the office. The routine was similar at every conference and Ron came to enjoy the redundancy. Knowing what to expect was comforting to him.

By 8:30 Ron had all of his material organized and began his routine conversations with each visitor who stopped for a look. Many left their business cards, which Ron tossed into an empty carton under the table, which would be half full by the end of the conference. He only held onto business cards from people with whom he seriously anticipated further contact or potential business opportunities.

Joe had come and gone again a few times by 11:00 a.m., but had finally settled into a conversation with some visitors at their booth. Ron decided it was time to take a break, so he went off in search of a cup of coffee. Walking through the main aisle he noticed how technologically-advanced some of the exhibitors' booths were compared to his own display of photos and models. Many had running movie clips filmed from drones depicting the design and construction of the labors; some even had 3-D booths which allowed you to roam through the simulated designs. Most of these were just gimmicks he thought. A truly talented architect simply needs a paper and pencil to create a masterpiece, or at least that's what one of his professors at the University of Rhode Island used to say.

Taking a quick look at the tour schedule that was posted near a water fountain, Ron noted that a city tour was to begin in fifteen minutes. Normally he and Joe would go together on such events, but it was only the first morning and someone should stay with the booth. So he decided to be spontaneous, something he usually was not, and join the tour.

There were already several people waiting at the meeting point in the lobby and as he approached a woman holding a clipboard came up to him and asked if he was interested in the tour. She then handed him a ticket. Apparently the tour was limited to the number of seats on the bus and it was first come, first served. There was complimentary coffee available, so he helped himself while he waited.

Casually eyeing the others, he realized that most were likely American or Canadian. He heard very few people speaking in other languages except for two women to his right who sounded as though they were speaking French. Both were smoking cigarettes and drinking coffee from cups that were heavily stained with dark lipstick. They were very petite and wore tight fitting skirts and equally tight blouses which revealed their almost complete absence of breasts. Not that he normally noticed those things but Ron couldn't help but smile at the difference between these two women and his Caroline, who he thought was so much more natural. But these two probably had the luxury of being cancer-free, unlike his beloved wife.

Taking a deep breath, he turned his attention away and was drawn by a conversation between the woman with the clipboard and a man. She was now speaking, in what was most likely the native language, Portuguese, to a man who Ron guessed was going to be the tour guide. The tours were normally given by local architects who were exceptionally knowledgeable about the history and design of the city. This man was about his own age, quite tall, with dark hair and medium-colored skin, dressed in casual black slacks and a beige polo shirt with black trim. The man nodded at the woman and she turned to face the group.

"May I have your attention, please?" she said.

She was primarily addressing the two French women who were the only ones still left in conversation, the rest of the group had obviously anticipated that the tour was about to start.

"We are ready to begin the city tour. If you will please follow me, our bus is waiting outside."

Two Japanese men were right on her heels as she led the crowd through the rotating doors. The men were simultaneously asking questions, which she was not able or willing to answer, and she gestured to the man with the black slacks and polo shirt. The two men hurried forward and were the first onto the bus where they took the front seats.

Ron was midway into the crowd as he entered the bus. There must have been about forty people in all. He took the first empty seat that he could find and said hello to the man next to him. He again took note of the others. The suspected tour guide was indeed in the seat behind the bus driver while the woman with the clipboard was in the seat next to him. The two Japanese men were across the aisle from them, busy in conversation, while the two *jeune filles* were in the back of the bus, which was where smoking was allowed.

Several people were attempting to open the windows due to the heat, apparently this would be an open-air tour rather than in over-cooled air conditioning.

Ron enjoyed the smell of the city, it was not like any of the cities that he knew back home. Providence had its distinct smells, like salt and decayed seaweed near the bay, oil near the large storage tanks along Interstate 95, or the aroma of oregano and fresh pasta in the Italian district on Federal Hill. Brasilia smelled fresh and raw. There was the unmistakable smell of vehicular pollution and the occasional waft of a sewage treatment plant, as well as a faint burnt odor to the air, but in general the aromas were subtle and mixed. It was as if the city was still trying to find itself and hadn't quite yet settled into its own scent.

After a few minutes the woman in the front stood up and motioned for their attention. "We don't have an audio system on this bus so you will all have to bear with us and try to keep private conversations to a minimum so that everyone can hear."

Heads popped up to look over the seats and the French ladies quieted down.

The woman continued, "I'd like to introduce myself. My name is Juanita Mendosa and I'm with the Brazilian Tourism

Commission," then, gesturing to the man beside her, she continued, "and this is Miguel Rodriguez."

The man in the black slacks and polo shirt then stood up next to her.

The woman added, "Miguel will begin by pointing out and describing locations of architectural significance to you today, while I'll be commenting on other sights of interest. Please ask either of us questions as they arise." She then took her seat while the man remained standing in the aisle.

"Hello. As Juanita has said, my name is Miguel Rodriguez and I'm with the firm New Horizons Limited, which is located here in the capital city, Brasilia. We primarily handle city projects, many of which I will point out to you along the way today." He paused while looking over the crowd. "I recognize a few of you from visits to our exhibit booth inside the conference hall this morning, and I hope that after this excursion, the rest of you will stop by to introduce yourselves and offer comments on our work... as long as they are positive, that is." He smiled and a few people laughed.

They pulled away from the curb and the tour guide began his planned talk. "Brasilia was constructed between 1956 and 1960, during the government of President Juscelino Kubitschek." Ron knew the basics so was only half paying attention as Miguel went on. "The city was planned for only 500,000 inhabitants, but it has grown much more than expected. Brasilia's total population, including several neighboring satellite cities, is now around three million inhabitants—"

"'Three million people?" A man in the third row yelled out. "Where are they all?"

Ron had also wondered why the streets looked nearly

empty. It was lunch hour and usually cities are teeming with pedestrians at that time who are searching for something to eat or trying to get some errands done.

The tour guide's answer was sensible. "Almost all of the office buildings have their own malls inside. You can get your lunch, a haircut, and even buy your groceries without having to leave the building. It's quite convenient."

Quite horrible, Ron thought. He loved going outside for a walk during his lunch hour. He couldn't imagine being cooped up all day.

A few minutes later the bus stopped and everyone got out to take a brief look at the Sanctuary of Dom Bosco Church. Ron picked up a brochure as he entered the modern-looking building and read that Dom Bosco, for whom the church was named, was an Italian saint and the founder of the Order of Salesians. In 1883, he apparently had a dream of a utopian paradise city, which would be built as the capital of a great nation. Based on this dream, the founders of Brasilia planned their new capital.

He must have been a pretty influential man, Ron thought, to have a country build their capital based on a mere dream.

But the church was breathtaking. He admired the walls inside the sanctuary, which were made of purple and blue vitral, a type of stained glass. He read that the vitral allowed sunlight to filter in, which then bathed the interior of the sanctuary in a blue haze. An enormous chandelier hung in the middle of the church, upon which light was also playing a tantalizing game. But just as he was being drawn to the chandelier for a closer look, the tour guides announced that they were ready to leave.

As they reboarded the bus, each passenger was handed a paper-bag lunch containing a wrapped sandwich, chips and a bottle of water. Ron was relieved since his stomach had embarrassingly started to growl while in the church. The snack would hold him over until they got back to the hotel.

Other sites of interest, which they viewed from inside the bus, included the TV Tower, the Church of Our Lady of Fatima, which was the first inaugurated building in Brasilia, and the square of Three Powers which included the Planalto Palace, the Federal Supreme Court, and the National Congress.

The tour took almost two hours and Ron felt a pang of guilt as he started back towards his display booth. As he approached, he saw a couple of men engaged in conversation with Joe. He paused for a moment and watched as Joe was displaying some sketches from a project they had completed last year.

The men seemed to be fascinated, constantly nodding in agreement with all that Joe had to say. Ron knew that his colleague was young and, outside of work, didn't seem to have his act together, but when it came to architecture he was a different man. He absolutely loved his work and that became clear when he spoke about it. He was passionate and intense but never condescending. He made everyone feel as though they could have come up with the same unique designs. He was brilliant with numbers and had completed a degree in civil engineering prior to studying architecture.

But his true gift was in drawing.

His mother, who was a professional artist, had enrolled him in art classes since he was five years old, and it pleased her to find that he actually enjoyed it. During his summer breaks from college, Joe decided to forgo working and instead enrolled in both art and design courses at the Rhode Island School of Design. Despite not having a degree in any of the artistic disciplines, he was clearly naturally gifted.

The two men handed Joe their business cards and shook hands as they departed. Ron approached the table to Joe's slight surprise.

"Hey! Where have you been?"

"Did you miss me?"

"Well, you left me to run the show on my own!"

"And from the looks of it, you're doing just fine." He nodded towards the two men who just left.

"Yeah, those two guys were from the Netherlands! You know, Holland? Windmills and stuff. I asked them if they could get me some of those wooden shoes! Ha! It's funny, the different people you meet at these things. So, did you eat?"

"Not really, I was on the city tour."

"What, without me?!?"

"Yes, well, I needed a little fresh air, so I hopped on the bus." He noticed that the crowd had diminished a bit. "So, you want to close up shop for an hour and grab something to eat?"

"No need to ask twice!" Joe grabbed his wallet from his briefcase, spun himself around the table and flashed his 'Cable Guy' grin.

Ron just rolled his eyes and shook his head. To be young again.

They passed through the lingering visitors towards the exit doors and Ron made note of some of the exhibitor's booths that he intended to stop by later in the conference. Close to the main doors he spotted the New Horizons emblem and thought the name looked familiar, and as they approached he recognized the man sitting behind the table as the tour guide.

He walked towards the table and turned to Joe, making sure that he noticed the detour. "Hey, come here a minute, I want to take a quick look." He pointed to the booth.

There were two others standing in front of the table looking at some drawings and, as Ron advanced, the tour guide sitting there apparently recognized him.

"Did you enjoy the tour of our city?" the guide asked.

"Yes, it was fascinating. Very different from home."

"Where's home?"

"Rhode Island." Ron replied, briefly forgetting that he was no longer in the States. Then he remembered that little RI may not be recognized worldwide so he added, "in the US."

"That's part of New York, right?" asked Miguel.

Joe began chuckling while Ron politely corrected the Brazilian. "No, you're probably thinking of Long Island."

"Long Island, it's that island in New York, right?"

"Yes, but we are from Rhode Island."

"Oh, sorry, then where is that?"

"Actually Rhode Island is not actually an island, it is the smallest of the fifty states, and we are about 200 miles east of New York City, nestled between Connecticut and Massachusetts."

"In New England?"

"Correct."

"The Colonial style is one of my favorites."

"Really?!?" Joe remarked, now joining into the conversation. "You couldn't get much further apart, between the modernism of Brasilia and the colonialism of New England."

"I suppose that's exactly why I find it so appealing." Miguel smiled, and turning to Ron he added. "Perhaps that's why you found our city so 'fascinating', I believe that was your choice of words. Might it be the extreme contrast between our familiar environment and someplace exotic?"

"I never thought of Rhode Island as being exotic!" Joe said.

"I have the same feeling about Brasilia." Miguel responded to prove his point.

Ron gestured to a model on the table. "This building looks familiar. Didn't we see this today?"

"Yes, actually we did. It's the new Sheraton Hotel and Conference Center adjacent to the government building sector. We finished the designs two years ago and construction was completed in January. It went up very fast."

Miguel went on to discuss its conceptual design, and intricate city approval process. They noted that their firms did similar work, although the styles varied greatly. Then, responding to questions from Joe, he discussed his formal education in Brazil and they noted the differences between the academic preparation in Brazil and that in the United States. Brazil required a two-year apprenticeship upon completion of schooling, whereas the US required more classroom hours. Miguel discussed his professional influences and showed them some sketches, which he had submitted for a historical museum that was still in the bidding stages.

Ron then heard a gurgling sound coming from Joe's stomach. "Hungry?"

Joe laughed. "Did you hear that?"

"I even heard that," Miguel said from the other side of the table. "You'd better put something in there."

"Any suggestions on local eateries?" Joe asked. "We've had both a dinner and breakfast here in the hotel and I'm ready to try someplace else."

"There is actually a great place just two blocks from here."

Miguel began describing the way to the restaurant when Ron suggested that he join them.

"Sure, why not. I haven't eaten lunch yet myself." He turned to his colleague and said a few words in Portuguese, then joined Ron and Joe, leading them out of the hotel.

"By the way," Ron said, "my name's Ron Stanley and this is my colleague Joe Pinochelli."

"Nice to meet you both, my name's Miguel," he said while shaking Joe's hand, who had not been on the tour. "Miguel Rodriguez. So, are you ready for some real Brazilian food?"

The conversation over lunch continued along the line of professional interests, then diverged to cultural differences between the United States and Brazil concerning food, climate, leisure activities and economic problems.

Ron looked at his watch and was startled at the time. "Wow, it's 4 p.m., we've been here almost two hours."

"There's another difference between our cultures," Miguel remarked. "We take our time over our meals and don't eat much fast food. In America you have, what do you call it... a 'powerful lunch'?"

"Power lunch." Joe corrected. "That's right, a three course meal in 20 minutes!"

"Three courses in twenty minutes?!?"

"Yep, Big Mac, super-sized fries, and Coke as the main

course, a Snickers for dessert and coffee to go *á la fin*." Joe gave his hand a flourish and pronounced *á la fin* with his best attempt at a French accent, while both Ron and Miguel rolled their eyes.

Miguel learned quickly, Ron thought.

"You have to work 8 hours a day with this guy? How can you bear it?" Miguel joked with Ron.

"He grows on you!" Ron said, giving Joe a punch on the arm.

They finally paid their bill and stepped out into the warm sunshine to make the short walk back to the conference center.

"That meal was fantastic. Thanks for the tip."

"My pleasure." Miguel held open the door to the hotel for Ron while Joe lingered before a placard on the sidewalk. "How long are you staying in Brasilia?"

"Just until Saturday morning. Unfortunately, because I would love to try a few more meals like the one we just had."

"Well, if you'd like, why don't you join me and my family for dinner tomorrow night. My wife is a terrific cook and, although her English is not good, I know she would enjoy meeting an American, or two. You could even bring your side-kick." Miguel raised his voice at the end and looked in the direction of Joe who was just approaching them.

"Side-kick? Me? Where are we going?" Joe asked.

"Miguel has invited us to dinner at his home tomorrow."

"Wow, sounds great. I'd love to see what the typical Brazilian home looks like."

"What do you mean *typical*?" Miguel said with a smile. "My house is a palace!"

"Of course, your highness. We'd be delighted to attend dinner at your residence." Joe mocked in his French accent again.

"Good, why don't you plan to ride home with me at the end of the sessions tomorrow. Meet me at my booth at about 5:30?"

"Great. See you then!" Ron said, slapping Joe on the back and giving him a tug in the direction of their exhibit. "In the meantime, let's get back to work!"

Chapter Fourteen

The precinct was buzzing as Nick entered the building on Thursday morning. He was early for a change and was glad not to have to deal with everybody's smart-ass comments as he made his way to his office.

He sat down with a cup of coffee and bit into the apple that Sonja sent with him. She always insisted he take at least one piece of fruit to work. When she first began this ritual, he had always forgotten that he had the fruit with him and, upon coming home, Sonja would find the piece still in his jacket pocket where she had tucked it that morning. After a few lectures about 'staying healthy for your family's sake', he realized it might do him some good to eat well, not to mention pacifying his wife. So he got into the habit of eating his fruit first thing in the morning, so that he wouldn't forget.

He began leafing through his in-box, and almost by intuition, sensed that his boss, Esposito, was standing in his doorway.

Without looking up from his paperwork, he said, "Morning, *Urso*. How's the day treating you so far?"

Taking a seat across from Nick and stretching out his long legs, Esposito said, "Not bad... but, then again, anything beats swallowing your tongue while strapped in a straitjacket."

Nick cocked his head in confusion and disgust, then tossed his apple core in the waste basket. "Come again?"

"You heard me right the first time, Nick."

"Shit, don't tell me you have another new case for me. This time some wacko up in Carandiru? Come on boss, I refuse to go up there again, that prison gives me the creeps." Esposito laughed and Nick relaxed a bit. "So, what is it?"

"It's not a new case. Well, at least that's the assumption I'm going on for now." He thought for a moment before continuing. "I think it may be related to that professor's death that you're looking into."

"How can a nutcase be related to—"

"I didn't say the guy in the straitjacket was a nutcase, you did. Actually he couldn't be more the opposite. Apparently he was a colleague of the late professor."

Esposito lay a brief report on the desk in front of Nick. There was a small photo clipped to it of a man that looked familiar.

"Isn't that—"

"A guy named Montagna." Esposito replied. "Professor Ricardo Montagna. Ring any bells?"

Nick arrived at Mandaqui Hospital at 9:30 a.m., having walked the few blocks from the precinct to avoid the late morning traffic rush. He identified himself to the nurse at the reception desk, and she began giving him directions to the hospital morgue. But having been there countless times before, Nick already knew the way, and gave her a quick wink as he hurried off.

Dr. Alexander Kasparov was on duty and Nick greeted him with a warm handshake. The two had worked together on multiple occasions and had always gotten along well. Dr. Kasparov had immigrated with his family to Brazil from Kazakhstan in the late 1980s to escape the turmoil preceding the collapse of the Soviet Union.

When Nick first met Dr. Kasparov, he thought that he was from Germany, since there were a lot of German immigrants in Sao Paulo, and the doctor had a thick European accent. But when asked, Kasparov hastily reputed the notion of being a "Westerner" and proudly told Nick stories about his beloved homeland.

Did you know that the Kazakhs were the first to domesticate the horse? Kasparov proclaimed. Can you imagine where all of humanity would be today without the work horse? For that, you can personally thank the Kazakhs. Nick had thanked him, personally, but said that he preferred motor vehicles. They'd both laughed.

When Dr. Kasparov first arrived in Brazil, it was very difficult for him to find work. After searching for something in Rio for two years and only able to occasionally land temporary work as a research technician, he was finally offered a staff position at Mandaqui and quickly relocated to Sao Paolo.

The doctor was a tall man with salt and pepper hair, fit and trim. He enjoyed the outdoors and the sea, and therefore, despite missing his homeland, he loved Brazil and its climate. Nick knew that Kasparov would never consider leaving now.

"Nick, to what do I owe this unexpected pleasure of your visit?" he asked with a genuine smile.

"I'll give you one guess."

"An interest in changing fields to forensic medicine?"

Nick exhaled with a chuckle. "Not this time, Alex. But who knows, I haven't hit my mid-life crisis. There may be hope for me yet."

"Then if I have a second guess, I'd say you are here to take a look at the professor I've got on ice."

"Yes, that's right. What can you tell me about him?"

Dr. Kasparov led Nick to the wall storage unit and pulled out the massive sliding tray numbered eleven. He zipped back the silver nylon bag and revealed a figure strapped into a straitjacket with the neck twisted slightly and deeply indented on one side. The mouth was wide open and there was massive bruising around the upper throat. "Is this your guy?"

"Yep, that's him," Nick confirmed. "Not a pretty sight is it?" he said, barely recognizing Professor Montagna from their brief contacts.

"Oh, I don't know Nick, I've seen much worse," the doctor replied. "I haven't touched him as you can probably tell. I called Inspector Esposito as soon as I got him down here."

"Good. Thanks," Nick said. "So, what happened?"

Dr. Kasparov zipped the bag back up and guided Nick to his office then took a seat. "Normally, this type of incident shouldn't have sent up any alarms. The guys from the psychiatric unit showed up to take him away and they found him dead. I was called up there to have a look, and when I asked for an ID, the nurse handed me his wallet where I found his university identification card. I did the autopsy on that other professor, Rodriguez, last week and I knew that the case was being investigated. Although medically speaking his case was cut and dry, ha, no pun intended," he slapped his knee and laughed at his own play on words, "so I thought I'd let you guys take a look at this before I cleaned him up."

"Thanks," Nick said, shaking his head at the attempted joke, "but I'm confused. What was he doing in the hospital in the first place, why the hell was he in a straitjacket and what does the psychiatric unit have to do with this case anyway?"

"Okay, let me start over." Dr. Kasparov poured himself coffee and offered some to Nick who refused. "Apparently the guy came into the ER on his own late yesterday afternoon. He began demanding a full examination, lab work, you name it. But the staff wouldn't treat him because he didn't appear to be in an emergency situation. Then the guy started to get hysterical, telling the nurse that if she didn't do what he said that she was soon going to have a corpse on her hands. He continued threatening her so she called for security and that's apparently when the professor grabbed her and started yelling that he was going to die. He lunged at two male nurses who came to subdue him and then began throwing everything in sight. He tried to fight off the security officers. They decided he was mentally disturbed with all the ranting and raving, so they put him in the jacket, locked him in an empty examining room and called the psychiatric unit."

Nick was taking a few notes, then looked up as the doctor paused. "Yeah? So why is he dead?"

"Well, from the disfigurement of his mouth and throat, it appears he swallowed his tongue."

"What? I thought that was a myth?"

"Actually, it is. But it sounds so gruesome that I still like to use the phrase."

Nick just looked at him with disbelief, then shook his head when he saw the doctor smile.

Dr. Kasparov continued. "Technically speaking, the tongue slips back against the pharynx, causing one to choke. Once the victim loses consciousness, well, normally, the tongue should relax and allow breathing to recommence, but it appears something else occurred here as well. Maybe heart failure from fright."

"From fright? What could have scared him... to death?"

"Being locked in a straitjacket and choking on your tongue wouldn't give you a little scare, Nick?"

"Enough to cause a heart attack?"

"Believe me, I've seen it all."

"I don't want to know." Nick was still taking notes and wanted to move on. "Were there any other marks on the body to indicate a struggle?"

"Like I said, I haven't touched him any further, in case you guys wanted to look at him first."

"Good idea." Nick thought for a minute. "But wasn't anyone around? See him struggling? How does this happen in a hospital?"

"Apparently, there was a bus accident shortly after that. The ER was full, and they sort of forgot about him."

"What about the guys from psychiatric? I thought they were coming to get him?"

"They've never been known for their quick response, you know that Nick. It was hours before they showed up, just as I was starting my shift."

Nick shut his notebook. "Alright, I'll need to talk to the attending nurse, the security team and the psychiatric staff who dealt with him, but first let's take another look at the body."

"My pleasure." Kasparov replied with enthusiasm and a broad smile.

"You've got a sick sense of humor, Alex."

"That's what keeps me going."

They both slipped on gloves before returning to drawer number eleven and then rolled the body onto a portable gurney. After wheeling it under the examination lights, the doctor drew back the zipper, slid the bag away from the corpse and then stood back waiting for Nick's lead.

He walked around the table pausing momentarily on each side. "How was he lying when they found him? On his back? Side?"

"He was on his back when I first saw him, but according to the nurses he'd been moved before I got there. The guys who found him said he was sort of unnaturally twisted up. They knew he was dead straight away, but they called in the nurses who laid him out, took vitals and confirmed death before they called me up."

"Can we take the jacket off him?" Nick asked next.

"Sure, give me a hand," Alex said, motioning for Nick to stand across the table from him.

They then rolled the body onto its side and unfastened the buckles binding the arm straps together behind the back. They released the four posterior buckles and slid it over the shoulders before laying the upper body back on the gurney. One more buckle freed the strap between the legs and the suit was slipped off. A few strategic cuts with a straight-edged razoi and Dr. Kasparov had removed most of the clothing as well.

And there lay Professor Montagna, stripped of his decency, his face blue and swollen, his neck disfigured, mouth gaping.

Nick wondered if the professor's death would forever quell answers concerning the demise of his colleague less than two weeks earlier. Or perhaps it would make it easier for him to find out the truth. If there was a murderer out there, two dead bodies would provide a lot more evidence than just one.

"Okay, doc, tell me what you see," Nick said.

"Well, if you look at the legs and torso, there is not a mark on them. Picture perfect. Although the guy looks as though he could have used a little more sun."

"Too late for that. So, what else?"

"That's about all I can see from the outside. I'll have to open him up to give you concrete answers. But, if I were to speculate, I'd say cardiac arrest was the most likely culprit."

"Most likely? Any other scenarios?"

"You know us scientists, Nick, we need to see the whole picture before drawing conclusions. Something had to have happened to induce this, perhaps a psychological or physical trauma, or chemicals. There are numerous possible scenarios."

"Right. So when can you work on this? I'd like to know as soon as possible what happened." Nick folded up his notes and slid them inside his jacket.

Alex looked at his watch. "I should have something for you by late this afternoon."

Nick shook the doctor's hand before leaving. "Thanks Alex, you know where to find me."

After returning to the precinct to pick up his car and his partner, Nick pulled up in front of Ricardo Montagne's house and parked behind an old two-toned station wagon. Nearing the front door they could hear the sound of a vacuum cleaner running inside. After knocking repeatedly on the door and then almost pounding, he and Tomas decided to wait patiently until the vacuum cleaner was turned off. They took a seat on the front step while they waited.

"This is the worst part of his job," Tomas said, "informing the family of a loved one's passing."

"Yep," Nick agreed. "That's why I brought you along," he said with a sly grin. But for some reason he hadn't expected Ricardo to have had a family, he seemed like a loner, but it appears he had misjudged him. He could only hope that it was just a wife and that there were no children at home. Kids love cops, but hearing from a police officer that their father is dead isn't exactly what they expect. And he didn't feel like acting as both investigator and psychotherapist at the moment.

The vacuum cleaner was finally turned off so Nick stood up and knocked on the door again. A moment later it was cautiously opened by a plump woman who appeared to be in her early fifties. She seemed very apprehensive. Nick pulled out his badge.

"I'm from the police. My name is da Silva and this is officer Castagna." He gestured to his partner. "Could we please have a word with you?"

"Senhor Montagne is not home. You will have to come back later."

"Actually, we're not looking for Senhor Montagne. Are

you his wife?"

"No, no. I do the housekeeping. Do you want to leave a message for him?"

"That won't be necessary. Actually, we have some bad news concerning Senhor Montagne. He unfortunately passed away last night."

"Oh, meu senhor!"

The woman appeared unsteady on her feet and grabbed hold of the door jam. Nick gently pushed open the door and offered to help. "Why don't you sit down for a moment, ma'am?" He helped her to the couch and then sat down in a chair opposite her.

"Are you alright? Can I get you a glass of water or something?"

"No. I'm fine. But senhor... What happened?"

"Well, I'd prefer not to discuss the details right now because we are conducting an investigation, but I can tell you that he died in the hospital."

She seemed to be carried away in thought, but Nick knew that he needed to get some information from this woman. "Do you know if Senhor Montagne has a family here? Wife, children, parents?"

"No, he did not have a wife or children, at least not here, he lived alone. But I do not know if he has parents or siblings that are living. We rarely spoke about personal things."

"I understand. Would you mind if I asked you a couple more questions Senhora...?"

"Gonzales. Yes, you may ask, but I don't believe that I will be able to help much."

Tomas took out a small notebook and began writing.

"When did you last see Senhor Montagne?"

"I...I...I do not remember. We almost never saw one another. I come during the day, while he is working. I clean and I put food in the refrigerator for him."

"Do you come on the weekends, too? Perhaps you saw him then?"

"No, I come only Monday through Friday, and not on holidays either. I have my own family to care for, too."

"Of course," Nick replied, trying to maintain a level of sensitivity. "Was Senhor Montagne a good man to work for? Did he treat you fairly?"

"Oh yes. He was very fair." She sighed apparently weighing the loss of future income. "I will miss this job."

She began wringing her hands and was apparently drifting off in thought again, so Nick realized he should wrap it up.

"Just one more question, Senhora Gonzales. Were you aware if Senhor Montagne was currently ill? Had he been at home sick at all recently when you were here?"

"No, no, he is never here. And I do not know if he was ill. But...", she paused in thought.

"But what?" Nick asked.

"Well, it was just a bit strange." Her eyes swept across the room. "When I arrived here today, things were a bit out of order."

"Really? How so?"

"Well, normally, everything is neat as a pin. But today his bed was unmade, there was food on the kitchen counter and there were papers all about in his study. Of course I made the bed and cleaned up the kitchen, but I thought it best that I leave everything as I found it in the office."

"That's fine, Senhora—"

"Oh!" She interjected as she remembered something else. "Actually, maybe Senhor Montagne was ill…"

"Why would you say that? Simply because things were out of order?"

"No, no. Because he has one of those medical bracelets. Sometimes he leaves it on his bureau and I saw it this morning while dusting."

"Do you know what his condition was?"

"No, I don't know. He looked very healthy to me. But the bracelet is still there, if that will help you. And there's a medicine cabinet in the bathroom, perhaps you'll find something there."

"Thank you, Senhora Gonzales. You have been quite helpful. Why don't you gather your things and head home now. I will lock the door when I leave."

"Yes, yes. I will do that."

She first put away the vacuum cleaner and other cleaning products, then gathered a small grey purse and a shopping bag from the kitchen. She nodded a goodbye on her way out.

As soon as the door shut behind her, Nick and Tomas began their search. Tomas went into the kitchen while Nick headed towards the hall leading to the bedrooms. The first door on the left was a bathroom, but the next door on the right was the master bedroom. The room, like the rest of the house, was very sparsely furnished and extremely orderly, so it took only a moment to spot the medical bracelet on the dresser.

As a precaution Nick decided to put gloves on before

touching anything else in the house, since one never knows where an investigation may lead.

He picked up the bracelet and turned it over. The words SEIZURE DISORDER were printed in large, bold letters.

Hmmm, that's interesting, Nick thought, although unfortunately it didn't set off any bells or whistles. But why did he not have it on him at all times? Maybe he had a second one or a neck chain with a metal info plate attached, but Nick couldn't recall seeing one in the morgue, and Alex would have surely mentioned it to him had there been one. The only conclusion he could come up with was that Ricardo must have had other things on his mind when he left the house. It was simply forgotten.

He dropped the bracelet into a plastic evidence bag and started toward the study.

As the housekeeper had mentioned there were papers strewn about. What she hadn't mentioned was that the computer was on with a screen saver flashing across the monitor. Some people leave their computers on 24/7, but Nick had a feeling that Ricardo didn't fit into that group. Something must have caused him to hurry off. But what?

Nick took a look at the papers on the desk and floor. Most were completely unintelligible to him, multiple hand-written series of numbers and symbols, but others were unmistakable, including a birth certificate and some dated medical files.

He picked up the birth certificate and wasn't surprised to see that it belonged to Professor Montagne himself. The medical files were also his own personal records. Maybe the guy was sick, or terminally ill even. But there were still too many unanswered questions.

Nick took a deep breath and sat down at the computer terminal, wondering if he'd have better luck with this machine than he did with Professor Rodriguez'. It had apparently gone into an idle mode with the screen saver dashing equations across its display.

He shook his head at his own ignorance. Had he known that detective work would require a pseudo-degree in computer science he may have chosen another field.

Slightly shifting the mouse, the screen saver disappeared and the background lit up to reveal a series of unintelligible lines of mixed numbers, symbols and letters. Nick cursed under his breath at again being outsmarted by a machine. He could work his way through Microsoft Windows well enough, but what he was now looking at was not the familiar user-friendly format that he was used to.

In front of him was what appeared to be a programming language run in DOS, which was about the best he could come up with.

He would have loved to try to find out more from the computer by opening up the email program, or searching through the file folders but there was no open toolbar on the screen and he didn't dare start pressing any buttons and risk losing what was displayed. Instead, he picked up his phone, pressed a number and waited.

"Valdez. Aiming to please."

Nick forced back his chuckle, then began in a deep commanding voice, "Officer Valdez, have you had your psychological exam required for employment?"

"What? Who's this? Nick?"

"That's right, Manuela," returning to his normal voice,

"and I'm shocked and hurt that you left me standing for our lunch date at the Ritz!"

"Was that today? Damn it, I knew I was forgetting something." She mocked a boo-hoo, then got serious. "Actually Nick, I'm glad you called. I've got something really good for you."

"That's what I was hoping for after our lunch date." He tried to force the image out of his head while saying it, but since she was one of the only women who would joke with him like that, he would take what he could get.

"Uh-huh, alright Nick, but you'll have to take a rain-check on that one. I've got the information you were looking for, actually my nephew has it, but I think you should come in here and we can go through it in person."

"Your nephew? Is he there? I was actually hoping to have him take a look at another computer."

"Nick, you've got to talk to this kid." Then she apparently turned to someone in her office and asked something in a muffled tone.

"Manuela, you still there?"

"Nick, he says he has some time. He's done with classes for today."

"What, is your nephew in high school or something?"

Almost choking on her sip of Coke, Manuela said, "Yeah, that's a good one Nick! No, my nephew teaches classes. Ramon is a graduate student just finishing up his doctorate at the university. And you might find it interesting to know that his supervising professor, with whom he has worked very closely for several years, just passed away: a man named Rodriguez."

"Oh, Manuela, you are a goddess."

"That's what they tell me. So, when can you get back to the office?"

"Actually, what are the chances that you can bring your nephew out here where I am? I've got another computer and a pile of notes that look like chicken-scratch to me. But I have a feeling they are the key to this case."

"Sure, give us directions and we'll be there as soon as my big old Chevy can carry us."

Was she referring to her car, or... ah, forget it.

Nick gave her the directions, then decided to see if Tomas was having any luck with the rest of the house.

Chapter Fifteen

"Mom! Mom! Hey, wake up!"

Caroline felt a tugging on her shoulder and heard the shouts as she was startled out of a sound sleep. Opening her eyes to the sight of her daughter standing by her bed, fully dressed and glowing with a giant smile, she sat up in a bit of alarm.

"What is it, honey? Everything alright?"

"Everything is great! Daddy comes homes today, remember?"

She didn't remember. She couldn't think straight.

What time is it? What day is it???

She looked at the clock to see that it was already 8 a.m. Why hadn't her alarm gone off? Maybe she forgot to set it, or could it be Saturday already?

She forced herself out of bed and pulled open the drapes allowing the sun to stream in. Then she began to think more clearly and realized her daughter had the days mixed up. "Sophie, today is Thursday. Daddy doesn't come home until Saturday. That's two days from now."

"Aw, mom. I want him to come home today."

"I would like that too, but we can't change his schedule.

The plane isn't flying home yet, so we have some more time for just you and me."

Caroline sat back down on her bed and pulled Sophie into her arms. "I love you," she said as she cradled her precious daughter in her arms.

Then, leaning back to look at her little girl, she said, "Now, I have to get moving quickly here. I must have forgotten to set my alarm, so it's great that you woke me up."

But she was still wondering how she'd managed to sleep so deeply. Then the abrupt reality of her illness took hold of her and she realized why she had overslept. Sleep was like a healing drug, freeing her from the burden of her illness, until she wakes up and the nightmare rushes back in.

She quickly showered and dressed, then went downstairs to find that Sophie had already let Max outside, and she was now standing on a chair at the kitchen counter trying to direct Cheerios into her bowl from a huge yellow box.

After helping her daughter with her breakfast and fixing coffee for herself, they got out the door and on the way to daycare by 8:45 a.m., just barely in time for Caroline to get herself to her doctor's appointment by 9:00 a.m.

"Good morning, Caroline." Dr. Mansfield offered a warm smile and handshake before pulling back a chair for her to sit on. "How are you doing?"

"Well, I miss my husband a bit." Caroline knew this was not the intended subject of the question but she wanted a little small talk before getting down to business.

"Miss him?" Then the puzzled look vanished, "Oh, that's right, you said he was going to Brazil this week. Have you spoken with him?"

"Yes, he is having a wonderful time. The conference is a huge success, with record attendance. He says that he has learned a lot from the seminars, and is absolutely enthralled with the city."

"Which city is he in? Rio de Janeiro?"

"No, in the capital, Brasilia." She had become so at ease with Doctor Mansfield, as if talking to her sister or close friend that she continued as if they were sitting over a cup of coffee. "He has even made a friend!"

"A friend! That sounds suspicious. Of the male or female variety?" she raised an eyebrow and grinned.

Caroline laughed. "Male. At least that's what he told me," she said in mock doubt. "Anyway, the man lives there in Brasilia with his wife and children. He is apparently an architect too and they have similar interests. Ron and his colleague Joe have even been invited to their new friend's house for dinner. I'm so curious to find out what their homes look like, what they eat... Oh!" She suddenly realized that she was just babbling away and taking up the doctor's valuable time. "Sorry, I got a little carried away there."

"How do you mean?"

"Well, you don't have all day to listen to me ramble on about my husband and his travels."

"Don't give it a second thought, Caroline. Anyway, there are plenty of magazines in the waiting room for the other patients to keep busy with!" She smiled and winked, causing Caroline to relax.

Dr. Mansfield then opened the medical file folder in front of her and looked up at Caroline with a more serious face. "Are you ready now to talk about you?"

"I'm ready," Caroline responded with confidence. She felt more emotionally stable than of late, and she thought her doctor sensed it too.

"I have Sophie's test results here," Dr. Mansfield began, "and unfortunately she is not a full match."

Caroline initially let out a sigh of relief knowing that she wouldn't have to put Sophie through any more testing. It was then that she was hit with the other implication of the results. No match meant no cure. Her heart plummeted. She had just been feeling so strong and now she was back on a low. It was like her emotions were on a roller-coaster ride.

"Caroline? Are you alright?" She was as white as a sheet, so Dr. Mansfield got up to pour her a glass of water. She handed it to her while kneeling down next to her chair. "Here, drink this."

In a low voice, "I'm fine, Doctor Mansfield", she replied.

Although it was clear to the doctor that she'd lost some of the confidence that she had just moments ago.

"Honestly," Caroline said, "I didn't expect her to be a match, and in a way I'm relieved that she doesn't have to endure the medical procedure." She closed her eyes and let out a deep sigh. "But on the other hand, having Sophie grow up without her mommy isn't going to be easy on her either."

"Hey, wait a minute, Caroline. We are not by any means finished here, I hope you are not giving up already?" She pulled a chair up beside Caroline and took her hand. "We still have the registry search, and we could find a match there. Also there is chemotherapy treatment, which could help."

"I know, but—"

"No *buts*, Caroline. You need to be mentally strong to fight this!"

"Okay, okay. There are just so many highs and lows in this. It is mentally and physically exhausting. But I guess you are right. I need to do everything possible, for Sophie's sake."

"Not just for Sophie's sake! For your own sake as well! You are still young and have the potential for wonderful experiences ahead of you. You need to want to beat this for *yourself*, not just for your family."

Caroline's chin quivered as though she were about to cry, but then took a deep breath and sat up straighter in her chair. "You are right, Doctor Mansfield. This is a difficult period for me. I sometimes feel like my emotions are all over the place. I was doing great a few minutes ago when we were talking about Ron, and now I'm about as low as it can get."

"That's normal."

"I'm trying to come to terms with all of this," Caroline said, "but you're right, I need to do this for me too. I don't want to die, at least not yet."

"That's better." Dr. Mansfield returned to her seat behind her desk before continuing. "So, are we ready to start the battle?"

Caroline wiped her nose with a tissue and forced a smile. "Ready, Captain."

They both laughed.

"Good. Now, on one hand, I think you should start on chemotherapy as soon as possible, so I've set up an appointment for you with Doctor Richard Blake for consultation."

"Wait a minute, do I really need another doctor? Can't you see me through this? And what do you mean 'on one hand'?"

"Caroline, I'm not an oncologist, you know that, and I want you to be handled by the best. Doctor Blake is the best in New England. He would oversee your chemotherapy treatments. In the meantime, I've put your data into the donor list to see if we can find a bone marrow match for you. But I'm not abandoning you, I'll be with you every step of the way, if that's what you want."

"Yes, I'd like that."

"So would I." Dr. Mansfield's face softened and she smiled. "Caroline, I like you, and I went to the same high school as your husband. I feel a bit of a personal attachment here and I give you my promise that I'll be there for you in every possible way that I can."

"Thanks, Doctor Mansfield. That means a lot to me." And it did.

"Good. Now, as I was saying, on one hand I'd recommend that you begin chemotherapy right away. That is the scientist in me talking, whereas the gambler in me has another strategy."

Caroline's curiosity was peaked. "Go on."

Dr. Mansfield closed her file folder and leaned back in her chair. She appeared to be trying very carefully to find the right words to continue with what she wanted to say next.

"What is it Doctor Mansfield? What's your other strategy?"

Dr. Mansfield took a deep breath before beginning. "Caroline, remember our discussion about you and Ron sharing two of the same HLA molecules?"

"Yes, of course. But you said that it wasn't enough for him to be my donor, right?"

"That's true. But I did go on to explain that the chances of your children being a perfect match are one in four."

Caroline wasn't exactly sure where this conversation was leading, since she thought the option with Sophie was already a closed subject. "Doctor Mansfield, you said that Sophie was not a match and can't be my donor, so I'm not sure what you mean."

"Let me try to put this more clearly. I am not talking about Ron or Sophie as being a donor for you, but instead... well... a second child."

"Doctor Mansfield, Sophie is our only child..." but then it clicked, and Caroline was speechless.

"Caroline, if you were to have another child, it could potentially save your life."

"But that is ridiculous. Are you suggesting that we create a baby to be used to cure me?"

"In essence, yes."

Caroline gasped for breath.

"But the timing is crucial," Dr. Mansfield continued. "Chemotherapy could cause you to become sterile, for this reason you would have to choose between immediately beginning chemotherapy or immediately trying to conceive a second child."

Thoughts were running a mile-a-minute through Caroline's head. Could this be true? Would it be possible? A brother or sister of Sophie to be used for medical purposes?

Without even thinking, the words came out of her mouth. "I couldn't do it."

"Caroline, it is something that needs to be considered."

"But I didn't even like the idea of using Sophie as a donor, how can I seriously consider bringing an innocent little baby into the world simply on the off-chance that it could be used to help me? That is so selfish."

"I understand, Caroline. It's not a simple issue. There are many factors to be considered here and, believe me, there are no easy answers. This has been an ethical issue which has come under heated debate in recent years."

"Debate?"

"Yes, primarily concerning stem cell research and pre-implantation diagnosis."

"I've heard of stem cells, but pre-implantation diagnosis, what's that?"

"Essentially, it is the evaluation of an embryo to determine whether or not it is 'suitable' to, well, cultivate."

Caroline began to cry. "Oh, that's sickening. Not to mention completely immoral. It sounds like a chapter in *Brave New World*."

"Yes, immoral, unethical, there are several contexts to which it has been corralled, but others have heralded this research as revolutionizing medicine."

Dr. Mansfield took a sheet of paper from the side of her desk.

"Let me read something to you Caroline. This is an excerpt from a study by the French National Ethics Committee concerning pre-implantation diagnosis: *In no case should a human being be conceived just for the purpose of using his own cells to cure his or her brother's or sister's disease. He or she should be conceived by his or her parents only as a moral end in his or herself-- in no case as an instrument.*"

"It sounds as if that closes the door on that topic," Caroline declared.

"Not exactly. First of all, the Committee does not create laws, it simply evaluates philosophical questions and issues opinions, which are not legally binding. This statement essentially represents your initial feelings on the subject."

"Yes. It does exactly," Caroline confirmed.

"Which is why I wanted to read it to you so that you understand that this topic has been thought through extremely thoroughly and from all aspects by many people. But, second of all, and more importantly, I believe that your situation is a bit different." Dr. Mansfield folded her hands on her desk and leaned slightly forward. "Caroline, I know that you and Ron are good people and love your daughter. I don't believe for one moment that you would throw away the life of a child simply for the chance of saving yourself, which is the only reason why I'm suggesting this alternative to you. Let me ask you, were the two of you planning on having more children?"

"We had always hoped to have at least two children, but now..."

"Then if this child was already planned for, and conceived in your heart, and if you would love it just as much as Sophie whether or not it would be a medical match for you, then where is the moral sin?"

"I just don't know."

"And one more thing, Caroline, if the child were a match, the blood cell donation could come directly from the umbilical cord. The baby would not be physically affected in any way."

"Well, that would certainly be one very positive aspect."

Then another frightening thought came to her. "But I think we have to be realistic as to the importance of timing here. I may not be alive a year from now, what if I don't conceive right away, or die with the child inside of me? Or shortly after it's born, whether it's a match or not? The idea of leaving Sophie motherless is burden enough. I'm not sure I could knowingly do that to a second child."

"There is a lot for you to consider right now. I suggest that you and Ron find some time alone and have a very long, very serious conversation. You could also look on the internet and research this topic yourself."

"Yes, I'm sure Ron would want to do that." Caroline sighed deeply.

Dr. Mansfield slid her chair back and came around to lean against the front of the desk. "Caroline, this is not going to be easy. I've said that before, and there are going to be some very difficult decisions to be made." Her mouth quivered for a moment as if she were fighting back a tear. "You know we recently lost my brother-in-law to this same disease and I constantly question myself as to whether I did everything I could. But it's too late now, even if I found the right answers today. It's just too late."

Caroline stood and was about to give a comforting remark when Dr. Mansfield continued. "But it's not too late for you. We have some time, and we have some options, so let's use all the weapons we have to fight this."

Caroline's heart wept for her doctor, and she felt for a moment as though she wanted to win this battle not just for herself and her family, but for Dr. Mansfield, too.

"You are right," Caroline said, "this is a battle, and it's not only me who could win by beating this disease."

"That's right, Caroline. Everyone wins."

Caroline stood up. "I'll give you a call on Monday, after I've had a chance to talk to Ron." She opened the office door.

The doctor breathed a sigh of relief. "You need to be strong Caroline. That is a very important step in treatment. The mind has much more power over our bodies than most people think."

"I'll try my best," Caroline replied with honesty.

"That's the right attitude. I think we are about to start on a winning streak."

"Thank you, doctor. I hope you're right," Caroline said before she left.

"So do I," Dr. Mansfield whispered as she turned back into her office. "So do I."

Chapter Sixteen

Nick was in the front yard when Manuela pulled up to Ricardo's house. Indeed, she was driving an old Chevrolet.

She stepped out of the car followed by a young man coming around from the other side.

"Hey, Nick. This is my nephew, Ramon." She turned around and let Ramon walk in front of her. "Ramon, this is Inspector da Silva."

"Hello Inspector. Nice to meet you." Ramon placed the file folders he was carrying under his arm to shake Nick's hand.

Nick couldn't help but smile to himself. He liked Ramon from the moment he laid eyes on him. He had an intelligent yet reserved demeanor, similar to his aunt, but was clearly blessed by not having inherited the same physical genetics that his aunt was burdened by.

"Thanks for helping us out, Ramon." He took a seat on the front steps as did Manuela and her nephew. "I understand if this is difficult for you, Ramon, considering you were a student of Professor Rodriguez."

"Yes, Professor Rodriguez was my mentor. It was, and still is, a shock to me knowing that he is gone. I admired him greatly, and I learned so much from him," he paused, pondering, "which may actually be to your benefit."

Nick smirked at the young man's cleverness. "You mean by interpreting the files?"

"Yes, that's what I mean," Ramon said while gesturing to the files on his lap. "Professor Rodriguez was a brilliant man. It appears that he was working on something very interesting here and on a topic which is actually quite surprising to me."

Nick wasn't sure where Ramon was headed so he kept his questions very general. "In what sense?"

"Well, first of all, this code that he has written diverges from his traditional interests."

"And why is that interesting?"

"It's interesting to me because I've never known him to work in this field before. He was passionate about his work and we had many heated discussions about his current studies and some of his extraneous interests, none of which forayed into growth and decay. That is Professor Montagne's field."

"Yes, well that brings us to another topic." Nick stood up and gestured to the house. "This is actually the house of Professor Ricardo Montagne."

Ramon now stood also, "Oh? What are we doing here? Is he helping you out too?"

"Well, I'm afraid I have some more bad news for you, Ramon."

He then explained with as little detail as possible the events surrounding Ricardo's demise.

"I can't believe this," Ramon said. "First Professor Rodriguez and then Professor Montagne and both within a week of each other. The odds of that must be miniscule." And then, almost to himself, "And what will happen with the department at the university?" He shook his head, then turned back to Nick with a serious look. "So, how can I help you?"

Both Ron and Joe agreed that the conference so far had been exceptional. The most interesting speakers were usually placed in the first and second days, and they'd both been lucky to catch some great presentations that morning. They had even found some time to peruse the display booths.

By 5:30 that afternoon they had put away most of their material and agreed to pack up the rest of it the next day. The last day was usually uneventful and they'd planned to spend most of it in the city. Ron needed to find a souvenir shop, so that he could bring something home for his girls. He couldn't leave without doing that. It was a priority.

But they still had the evening ahead of them and Joe and Ron were really looking forward to spending it with their new friend and his family. As they were heading over to Miguel's booth, they saw him coming towards them.

"So, are you two ready for the experience of your lives?" Miguel asked with a smile as they walked through the hotel entrance and out onto the glistening sidewalk.

"I thought we were going for dinner!" Joe countered.

"We are," Miguel answered, "but one like no other."

"Sounds great. So, where are we headed?" Ron asked.

"To my car, first of all, then on to a traditional Brazilian dinner compliments of my lovely wife." Miguel led them around the corner and into a parking garage where they got into his car. "I know what she's got planned for you, and if I'd known that she was going to cook like that for American visitors, I would've had you over every night this week!"

Miguel drove them through the north-west route out of

the city along Via S1 Oeste, past the immense Parque Recreativo de Brasília Rogério Pithon Farias, then by the JK Memorial.

Joe asked why there was a statue of John Kennedy in Brasilia, to which Miguel explained that it was not for JFK rather it was built to honor Juscelino Kubitschek, one of the founders of Brasilia.

Then they turned onto Monumental Highway and shortly thereafter entered the satellite town called Cruzeiro Velho. The town itself was surrounded by green fields, but upon entering the district, it was wall-to-wall housing with very small gardens.

"Uh, nice area you live in Miguel." Joe tried to sound genuine but couldn't mask what appeared to be surprise at the lifestyle differences between Brasilia and his hometown in Rhode Island.

"You think so, Joe?" Ron said with a chuckle, knowing full well that Joe was having a cultural experience.

Miguel seemed to catch on. "You think this is bad? We used to live in the Quadras, an apartment in one of the block buildings along Exio Rodoviario Norte. It was loud and small and we didn't even have a balcony. Once my wife, Katie, got pregnant we decided we needed to get out of the city, so we ended up here. But not forever, I'm a Paulista, so one day I'd like to go home."

"You're from Sao Paulo?" asked Ron surprised.

"Yes, my family is still there."

Just then they pulled into a driveway and were greeted by three small children scrambling up to their father's car window.

Joe and Ron got out of the car and watched as the kids drew back to observe them.

Tugging on her father's pant leg, a little girl who looked to be about four years old asked, "Pai, são estes os americanos?"

"Yes, this is Ron and Joe. They are from the United States of America."

"Oohhhh..." she cooed as if awestruck.

Ron and Joe couldn't help but laugh. Then they were introduced to Katie, a very attractive woman with medium colored skin and long, wavy dark hair. She had a warm smile and greeted her visitors with a friendly handshake. "Welcome to ours home."

"*OUR* home," Miguel gently corrected her with a smile.

"Yes, our home," she repeated.

They entered the house and Ron was surprised at how similar it appeared to any home in the United States. He wasn't sure what to expect, but was relieved to see that they had essentially the same comforts that he was used to. The living room was furnished with an overstuffed couch and two arm chairs, soft sunset orange curtains hung in the windows, a television on the far wall, and a shelf full of kids' games and books was arranged in the corner.

"Something smells delicious!" Joe breathed in deeply a couple of times, all the while rubbing his stomach.

"You don't have to inhale the food here Joe, we'll sit down in a minute and serve it to you on a plate!" Miguel laughed, and the rest joined in.

There was a small formal dining area where they sat to eat. Apparently, the kids had already had their supper and were playing card games in the living room, so it was just the four of them.

Katie brought out plate after plate of wonderfully aromatic dishes, naming each as she placed it in the middle of the table so that everyone could help themselves.

There was feijoada, which was a stew made of black beans and pork, a chicken dish called frango ensopado, yellow rice, a mixed salad called salpicão and a basket of cheese rolls she called pao de queijo.

Miguel poured red wine for himself, Joe and Ron.

"You don't drink wine?" Joe asked Katie who was sipping water from a wine glass.

"Not now." She smiled.

"Katie's pregnant," Miguel said, his face also covered in a proud smile to match his wife's.

"Well then, let's toast to that!" Ron raised his glass and after a collective ringing of the crystal followed by a drink, they began filling their plates.

The three men were busy eating but Katie ate only sparingly as she had questions for her visitors. "Do you have children?"

Since Joe's mouth was full to overflowing, Ron answered that one. "Joe is still looking for a woman who will have him, so until he finds her, he'll be childless." Ron caught a kick from under the table and a side-long glance from Joe who was still trying to swallow his mouthful of food as Ron continued, "I have one daughter, named Sophie, she is four years old...and beautiful like her mother."

Katie beamed at his response.

"What is your wife called? Her name?"

"Her name is Caroline."

"Lovely," Katie said.

"Yes, she most certainly is!" Ron agreed.

Miguel and Joe exchanged a look and laughed. "Okay, you two, let's get back to eating." Miguel handed Ron the basket of bread.

Joe kept back his typical wise cracks and changed the subject to their shared work interests. "Would you like to hear about some of our current projects?"

"I'd be quite interested, yes." Miguel responded as did Katie with a nod.

Then Ron took a pen from his shirt pocket and began drawing on a napkin before Katie brought him a pad of paper. He sketched not only the buildings but the surrounding landscaping, roads and even whimsical people on the streets.

The dinner continued with interspersed antics from Ron making comments about Joe's artwork, which were always received with laughter from their hosts.

Katie had just finished clearing the table with the help of her two older boys and had brought in the homemade flan when the phone rang. She picked it up and gave a warm greeting to the caller before changing her expression to that of concern and handing the phone to her husband. Miguel walked out of the room as Katie returned to the table.

"Is everything alright?" Ron asked.

"Oh, yes. Miguel's sister is on the telephone."

But Ron could hear Miguel in the other room, and even though he didn't understand the Portuguese, he could tell that Miguel was trying to calm someone down.

Chapter Seventeen

Ramon needed a few hours to review all the paperwork found in Ricardo's home office. In the meantime, Nick had called in a team of field officers from the department to collect evidence and clear the scene. Tomas was leading up their efforts while Nick was focusing on Ramon's work in the office.

"This is phenomenal." Ramon appeared to be half talking to himself and half to Nick.

"What's phenomenal, Ramon?" Nick asked.

Gesturing to the stack of papers on the desk and the computer screen, Ramon answered, "I've reviewed these notes and compared them to what's on the computer here, which is actually the same model that was found on Professor Rodriguez' computer. I'm having trouble believing that this model can actually be possible. It appears to be logical but defies all laws of reason."

Ramon then looked at Nick and his eyes seemed to clear as though he just realized he was talking to a police officer rather than a fellow mathematician. "Sorry, Inspector, I don't mean to be so obtuse, but I am simply overwhelmed by what I'm seeing here. It seems as though Professor Rodriguez has developed an incredibly detailed growth analysis, which simultaneously attributes the effects of decline."

Nick, not being known for his patience, cut straight to the core. "Ramon, do me a favor, pretend I'm a four-year-old and try that again."

So Ramon took another approach to explain, this time using the Carbon 14 analogy. "Look at it this way, Inspector. Every living organism, from plankton to elephants has a rate of growth that can be mathematically predicted."

"Really?" Nick was amused at Ramon's attempt at explaining.

"Yes, really. And some materials, such as Carbon 14, have a rate of decay that can be mathematically predicted."

"Okay, so what does that have to do with the Professor's files?"

"Quite simply, it appears that the Professor has put growth and decay or decline in the same equation, allowing us to calculate *beforehand* the age of a specimen at the end of its lifespan."

Nick ground his teeth in frustration thinking that this lesson in biology was a complete waste of his time. "So, are we back to the carbon again? What type of 'specimens' are we talking about? Plankton, elephants?"

"Oh, no, Inspector, this is much more explosive."

"Then, what is it?"

"Well, that was my question, too. The equation is too complex to decipher its intent without a more detailed explanation. And that's where the *Death.doc* file written by Professor Rodriguez comes in."

"You got it decoded already?"

Manuela suddenly chimed in. "See, I told you he was a genius. It's in the blood." She'd been there the whole time and had brought a laptop computer with some of the relevant files to the case, including the *Death.doc* file, which Ramon had cracked in a matter of minutes.

"Yes, the Professor taught me how to code and, naturally, decode files, so it was quite simple."

"And what did you find?"

"It's a personal letter, from the professor to his wife." Manuela handed Nick her laptop so that he could take a look. "See for yourself."

Nick read about half-way through the page when he stopped and looked at Ramon. "This can't be possible."

Ramon smiled, reveling in the fact that a few moments ago those were his same words. "Inspector, I knew Professor Rodriguez was brilliant, and if I thought anyone could do something like this, I would have granted him the honor, but I never imagined..."

"But, how?"

"I haven't quite understood myself yet exactly how, but..." clearly excited now, Ramon continued, "it appears that Professor Rodriguez has developed a formula which can determine individual human lifespan."

"Or stated another way," Nick added, "predict someone's *exact date of death.*"

With a confirming nod, Ramon replied, "Exactly."

"But that's not possible, I mean, I could be hit by a car when I walk out of this house. How could that possibly be predictable?" Nick wasn't buying it.

"That's the genius part, Inspector." Ramon lifted the stack

of papers holding the equation. "There is an extremely complex risk factor here, which essentially dominates the equation."

"A 'risk factor'?"

"That's right. Everyone has an amount of risk in their lives, some people more than others, and if the right factors are put into this equation then even our susceptibility to risk can be determined. Or at least that's what Professor Rodriguez thought."

"Wow," Manuela muttered, clearly in awe.

Interrupting Nick's moment of shock was the shrill ring of his cell phone. "Da Silva," he answered with his eyes still locked onto the computer screen before him.

"Nick, it's Alex. Got a minute?"

Falling out of his reverie and back into reality, "Yes, of course, Alex, actually I wanted to get in touch with you. I found out that your case study in the straitjacket was an epileptic."

"Well, then, I guess this call is in vain. That's exactly what I wanted to inform you of. A common seizure caused his problems. He died of asphyxiation and since he was in the jacket and locked in a room, he had no chance of assistance."

"Well, from what I've just learned Alex, I think the guy's proverbial 'time was up' anyway."

Chapter Eighteen

After finishing up at Professor Montagne's place, Nick took Ramon with him over to the Rodriguez home while Tomas got a ride with Manuela back to the precinct.

Nick had called from the car to announce their visit and to make sure that Maria could be there too. By chance, she already was.

Upon hearing the news, Rosa shook her head, "I can't believe he is gone too. I just saw him the other day. How can this be?"

Maria stood close and had an arm wrapped around her mother's shoulder.

Rosa was clearly shaken by the news. "That will be a double burden on the math department at the university." She then told her guests about Ricardo's visit to her on Monday, and about his use of Eduardo's computer to try to salvage files for the students.

Ramon turned to Nick, "That must have been where he got the executable version of the equation."

Nick nodded in agreement.

"Thank you for stopping by personally to let us know about Ricardo, Inspector," Rosa said. They were all still standing in the entranceway and Rosa thought that that was all they'd come by to tell her.

Placing his hands in his pockets, Nick spoke very slowly, knowing what she was soon to learn may be very difficult to handle emotionally, "But, Senhora, there is more, much more."

"Oh, dear, that sounds serious. Then let me make some coffee and we can all sit down." She ushered them into the sitting room while she and Maria went to prepare the coffee.

"How do you think she'll react?" Nick asked.

Ramon smiled. "I don't know her all that well, but if she is anything like her husband, well, let's just say I think she'll be intrigued."

Rosa reappeared and poured them each a cup of coffee. After she made herself comfortable, they began to explain how they believed that Ricardo had used a mathematical equation to calculate the exact moment of his own death, and that he brought himself to the hospital in the hope of getting help, but that it was to no avail.

"Oh, poor Ricardo. How sad." Then she added, "But an equation to predict his own death? Could this be possible?" Rosa asked.

"So, you really know nothing about it?" Nick asked.

"I don't understand," Rosa answered. "Why should I know something about it? My husband didn't burden me with all the research being done by his colleagues at the university. Thank goodness!"

Nick then cautiously answered. "Senhora Rodriguez, the equation was actually developed by your husband, and we think that he used it on himself first to determine his own death."

"No, this can't be!" Maria exclaimed.

"I don't believe it, either," Rosa concurred. "That is something he definitely would have told me about."

Nick then produced the letter, which they had printed out on Ricardo's printer, just before leaving. They explained that it was written to Rosa by her husband just prior to his death and that they had discovered it on his computer at the University.

Although it was addressed solely to her, she read it aloud so that Maria could also hear.

My dearest Rosa,

During our lives together I was never able to keep a secret from you--not even when it came to your Christmas gifts--so I should not be surprised that even after my death you have found me out again. Thus, this letter could only have been discovered along with my research, the lifespan equation, and therefore I now owe you some explanations.

First and foremost, concerning the surroundings of my death, yes, I did know that I was going to die, and precisely when. I had tested the equation as best as possible on others who had already passed, and I found it to be extraordinarily accurate—amazing even myself! And thus I prepared for my own death as best I could, making sure that you and the children would not be overly burdened with paperwork and finances, and I peacefully went to God when he called. The reason that I did not discuss this with you nor the children, if not already clear, was because I knew it would bring you pain and that it would be difficult enough once I was gone; in effect I wanted to reduce the sum total of your suffering.

Rosa interrupted her reading, "Sum total! Sum total! Goodness was he ever a mathematician!" She was in tears clearly brought on by the joy of reading his words and the startling revelation of his unimaginable lifespan equation.

I hope you can understand and accept this decision; it was made purely out of love and respect for you and the children.

In terms of the lifespan equation, I was not so decisive. I could have destroyed it myself but wondered if it could somehow be used in a positive way. This decision I will leave to you. You have experienced first-hand what it is like to lose someone who had applied the mysteries of the equation to his own life and seen it fulfilled. You may want to have it immediately destroyed, or you may want to give it some thought, perhaps discuss it with Miguel and Maria, before making your decision. I have only one request. If you should decide to allow the equation to live, that you turn it over to my graduate student (and friend) at the university, Ramon Santos, so that he may control its use. I believe that he would be an excellent guardian of it.

Glancing over at Ramon, Nick then realized his relationship to the Professor had been very close, almost like father and son, and that he was likely suffering as much as the two women over the loss.

And last of all Rosa, I would like to reassure you that I died in peace. My years were full. My family, work and friends were always at balance in my life. God had truly blessed me, providing more than I could have ever hoped. You were a wonderful wife and mother and I hope that you continue as always, to be there for the children and grandchildren, to provide them with comfort, care and advice, and the simple pleasure of seeing your smile.

With all my love until we meet again,

Your Eduardo

"Oh, Father!" Maria covered her face with her hands and began to weep as her mother took her in her arms.

"Thank you for delivering this, Inspector," Rosa said as she released her daughter. "And to you, Senhor Santos."

"Please call me Ramon, as Professor Rodriguez did."

"Eduardo thought very highly of you Ramon, and not just because of your mathematical talents, he really enjoyed working with you."

"And I, him," Ramon answered.

At that, the two men stood to go. Rosa said goodbye in the sitting room, while Maria, having regained her composure, escorted their visitors to the door.

"I think we will be seeing each other again," Maria said to both the Inspector and Ramon, but Nick was already a few steps down the path so Ramon turned to Maria, and she noticed his surprise and a slight blush on his cheeks. "I mean, because of the equation," she said in clarification, "once my mother makes a decision, she will surely want to talk to you."

Ramon cleared his throat, "Oh, yes, of course," he'd nearly forgotten, "please call me any time to discuss it."

Maria gave him a shy smile, "Thank you, Ramon."

Returning the smile, he answered, "My pleasure, Maria."

Miguel couldn't make any sense of what his sister was trying to tell him on the phone. "Slow down, Maria, slow down. I can't understand you."

He tried without success to get her to calm down and start over from the beginning, but she was so excited that he finally gave up and insisted on talking to his mother who he had heard in the background.

"Mother, what is it? What's happened?"

Rosa explained about their visitors who had just left, the news she had received about the death of her husband's colleague and the letter from her husband.

"Are you alright, mother?"

"Do you know something, Miguel? I feel much better now."

"How so?"

"This letter from your father, it has washed away my fears. I feared that he suffered. Somehow, I feared that he should feel his life incomplete. But I was wrong, your father went peacefully, and he accepted it without question."

"I wish I was there to give you a hug. You're wonderful," Miguel remarked to his mother.

"Thanks, Miguel, so are you. I'll call you once I've had some time to think about what to do with this equation."

"I think you've already made your decision, mother."

Warming to the thought of how well her son knew her, she softly answered, "I think you are right my son."

Miguel returned to the dining room and appeared somewhat bemused. Sitting down in his chair, his wife took his hand in hers. "Is everything alright?"

Shaking his head, but with a sly smile on his face, he said, "You are not going to believe what my father has done."

"Your father? But..." Katie was confused, as were Joe and Ron.

Then Miguel decided to bring everyone up to date. Addressing Joe and Ron, Miguel began, "My father was a mathematician. He died two weeks ago."

"Oh, gosh, I'm sorry," Ron and Joe both said almost simultaneously.

"Yes, thank you," Miguel continued, "it was very unexpected. He was only 58 years old."

The same sly smile was still on Miguel's face, and he was shaking his head back and forth slightly in disbelief, which had Katie and his guests beguiled.

"What am I missing here?" Joe asked. "Why are you smiling?"

"Apparently," Miguel said, "my father had developed an equation which can determine how long people will live."

Ron tried to clarify. "You mean, the average life expectancy? I thought that was simple statistics."

"No, I don't mean *average* anything. From what I've just learned, he developed an equation that could be applied to any given individual, and if all the inputs are correct, then the exact date of death can be determined. For anyone and everyone on the planet."

The room was quiet for a moment, then Katie, looking back and forth between her husband and the astonished guests, broke the silence, "Eu não compreendo."

Miguel explained once again in Portuguese to his wife, who then sat with the same shocked expression as the rest of them.

"Well, Miguel, you were right about one thing," Joe remarked.

"What's that, Joe?"

"You said tonight would be an 'experience of our lives'!" He raised his glass. "So, let's toast to the experience of our now pre-determined lives!"

They lifted their glasses but before anyone could take a sip all four broke out laughing.

Chapter Nineteen

"Hello?" Caroline picked up the phone on the first ring in the hopes of answering it before Sophie, but apparently one ring was all it took for her daughter who got to it simultaneously.

"Hi, Caroline, it's me."

"Daddy!"

"Oh hi, Sophie, I thought it was your mother who answered the phone."

"Nope, it's me, but she's here too, I think," Sophie responded happily.

"Yes, I'm here too," Caroline chimed in, "how are you?"

"I'm good, very good. How are you?"

"Daddy are you coming home today? I want to go swimming."

"Not today sweetheart, I'll be home tomorrow and then we'll have all day Sunday to do whatever you want."

"Swimming. Can we go?"

"If you want. Now can I talk to your mother for a minute? I'll see you tomorrow. Okay?"

"Okay, daddy, bye!" But she hung up and was gone even before she had a chance to hear her father's goodbye.

"So, darling, how are you? How was your appointment with Doctor Mansfield yesterday?"

"Oh, it was interesting."

"Interesting? Is Sophie a match?"

"No, unfortunately not, but Doctor Mansfield has a creative suggestion for us. We need to talk about it when you get home."

"Alright. We'll talk tomorrow night after Sophie is in bed."

"Oh, and I have an appointment with an oncologist next week, on Monday, I'd like you to come with me."

"Of course. I'll be there every step of the way. But will Doctor Mansfield still be involved? I know you feel really comfortable with her."

"Well, Doctor Mansfield has a lot of experience with this type of cancer, but mostly from a personal side after losing her brother-in-law. She has done considerable research in the area on treatment, but is not a chemotherapy specialist."

"Chemotherapy? Did she suggest you start that right away? This is all moving so fast. I'm sorry that I'm not there with you."

"We'll have to talk about this when you get home. I'm fine, so don't feel guilty about not being here. I know you are enjoying yourself and that makes me happy."

"I love you, Caroline."

"I love you, too. I'll see you tomorrow night."

As if scripted, they each disconnected and momentarily closed their eyes, taking a deep breath before putting down their respective phones.

As Ron entered the conference center on the last day and

arrived at their booth, he was happy to see that Joe was just about finished packing up their gear. "Already packed?"

"Yep, most of the displays were already taken down last night, so I just cleaned ours up so we can check out the city."

"Good. You want some help getting these to the shipping office?"

"Nah, I'm all set, why don't you grab some breakfast and then we'll hit the town!"

"Joe, it's nine o'clock in the morning, we're not 'hitting the town'."

"You never know, this ain't little ol' Rhode Island!"

Ron just laughed as he headed towards the breakfast room. But passing through the display tables he decided to see if Miguel was there yet. Sure enough, he was busy packing up his own display.

Miguel's back was turned to Ron as he approached.

"Bom dia, senhor." Ron attempted his best Brazilian accent with one of the few phrases that he had learned that week.

Turning, Miguel began to reply before seeing who was addressing him. "Bom dia..., oh, hey, good morning, Ron." They shook hands. "Everything alright? You look a little tired."

Ron thought he looked fine, but his conversation with Caroline and the thought of her going into chemotherapy was heavy on his mind.

"Yes, I slept fine. I just got off the phone with my wife and daughter."

"And is everything alright at home? I'm sure they are looking forward to seeing you tomorrow."

Ron weighed whether or not he should mention Caroline's

illness to Miguel, then decided there was no reason not to. "Yes, Caroline and Sophie are really looking forward to my coming home, but...well, you see, Caroline is ill."

"Ill? A flu or something? Someone always seems to have a cold in our house, despite the heat here."

Ron shifted his feet and looked out towards the front doors, the warm sunshine streaming in suddenly giving him strength. "No, unfortunately it's not something as minor as the flu. Caroline has cancer. We just recently found out. She may have to start on chemotherapy as early as next week."

"Ó Deus! Ron, I'm sorry to hear about that."

"Yes, thanks, Miguel. She'll be alright. At least that's what I hope. The medical developments are quite advanced now."

"And you are lucky you live in America where she will get the best treatment possible."

"I guess you're right."

"But hey, Katie and I were talking after you left. We really had a good time last night and we'd like to keep in touch. Katie wanted me to be sure to invite to you to come back and visit sometime with Caroline and Sophie. Maybe Sophie can teach some English to our little Rosie."

"We'd love that, Miguel. I'll talk to Caroline about it...seriously, I will." He then added, "That might be just what she needs once she's done with chemotherapy, to get her mind off things and get well again."

"You may be exactly right, my friend."

After exchanging addresses, emails, and phone numbers, they shook hands and parted. But Brasilia was a long way from Rhode Island. And despite the best of intentions, Ron wasn't so sure if they'd ever see each other again.

Chapter Twenty

The small commuter jet from Boston touched down at Providence's Green Airport at 6:15 p.m. Joe and Ron were both tired, but adrenaline kept them going, as both were looking forward to the homecoming. They arrived at baggage claim just as the luggage carousel began to turn.

"So, you've got some big plans for tonight then, Joe? Taking Suzie out to dinner?"

"That's the plan. I sent her a few photos this week but I didn't hear too much back from her." He let out a deep breath. "It feels like we've been gone an eternity. I hope she hasn't forgotten me."

Ron smiled as he noticed that Joe actually appeared a bit concerned. He laughed, "Forget you? If you haven't yet noticed Joe, you're not the forgettable type. I'm sure she's sitting by the phone, as we speak, waiting for it to ring!"

"Nah…" Joe said. "Hopefully she's not that desperate! I have a reputation to uphold, you know. But seriously, I will give her a call. She said she doesn't like texting and I guess she meant it." Then he reached down and snapped up his suitcase with his left-hand and almost simultaneously grabbed Ron's bag with his right hand. "Got to help out the elders, you know."

Ron shook his head and they both laughed as they headed for the parking lot.

It was already getting dark as they drove along I-95 heading south. The two were quiet, both tired, but deep in thought concerning two women, for two very different reasons.

Then Joe broke the silence. "Is there anything I can do to help Caroline?"

Ron's opened his mouth but then wasn't sure how to answer, so Joe continued. "I mean, if she needs a donor or something. I could give blood or bone marrow, or..." Then he drifted off mid-sentence.

Poor Joe, Ron thought. He really wanted to help but was having trouble getting the words out. "Thanks Joe, I appreciate it. Actually, that would be great if you could give a sample for the registry. You never know."

Joe seemed to immediately perk up.

"Alright! Where do I go?"

Ron explained the details of where to go and what to expect. He knew that much from his hours of internet searching.

Twenty-five minutes later they arrived at Ron's house. Caroline and Sophie were waiting outside, tipped off by his phone call, and Ron stepped out of Joe's car to a warm greeting. Joe got out too for a second to say hello.

After she had given a huge hug to her dad, Sophie turned to Joe and said, "Hey, I know you!"

"Hey, I know you too!" Joe returned. "Aren't you Big Bird?"

Sophie squealed in delight. "NO! I'm not Big Bird. I'm Sophie!"

Kneeling down in front of her, Joe tousled his fingers

through her hair. "Hey, you're right. No feathers. You can't possibly be a bird!" He then took her up in his arms and walked over to where Caroline and Ron were watching them.

"Hi, Caroline. How are you doing?" Joe asked.

"I'm much better now that my family is reunited." She took hold of Ron's arm.

"I can imagine," Joe said. "Well, I won't hold up the homecoming any longer. Sounds like someone else is eager to see you, Ron." Joe indicated the barking coming from inside the front door.

"That's Max," Sophie declared.

"Just a minute, Max," Ron yelled in the direction of the house.

Joe placed Sophie on the ground then tossed his car keys in the air with his right hand and deftly caught them behind Sophie's back with his left hand. The trick was rewarded with more squeals of delight from the little girl.

"Joe's got a homecoming of his own this evening." Ron eyed Caroline and winked.

"Oh, nice. Have a great time, Joe," Caroline said.

"Behave yourself!" Ron called out as Joe slipped into the car.

"But what fun would that be?!?" Joe replied through his open window while backing out of the driveway.

They waved and Joe was off.

Ron grabbed his bag, took Sophie on the other arm, and followed his wife inside the house.

It was good to be home. As Joe had said earlier, it felt like they had been gone a lot longer than five days. Everything seemed different. No longer was he afraid to talk about Caroline's illness and her future, their future. He was ready to deal with it head-on, and use all of the resources available.

He had always known how good his life was, and how blessed he had been, but now he wanted to begin a new commitment of appreciating each moment. Somehow this trip to Brazil had been a reawakening for him, and he was looking from a new perspective at life and what was in store for him and his family.

The evening waiting for him had been perfectly choreographed by his daughter. The first thing he noticed after coming inside was a large 'Welcome Home' sign hanging over the dining table. Sophie proudly announced that she had colored it in herself. As if that wasn't clear enough. He was then informed by his four-year-old that she had cooked dinner for him and that he needed to take his seat at the table.

Sophie then hurried into the kitchen with her mom and moments later they brought out the food. There was ricotta-stuffed cannelloni, crispy garlic bread and Caesar salad. It was Ron's favorite meal.

After dinner Ron sat his wife and daughter on the couch while they opened the gifts he had bought in Brasilia.

It took a few minutes for Sophie to figure out how to work her new little percussion instrument. It was a small drum with a handle. On either side of the instrument were two small wooden balls attached to leather straps about three inches long. Finally, when Sophie's frustration was at a hilt, Ron demonstrated as the clerk in the gift shop had shown him. With a quick snap of the wrist the drum rotated back and forth while the balls rapped either side of the taught leather in synchrony. Sophie snatched the gift out of her father's hand and tried it herself, with little success, but was happily distracted when Caroline unwrapped a gourmet box of fine chocolates and coffee.

"Mmmmm...the coffee smells wonderful," she remarked. And at about the same time five little fingers tried to reach for the chocolate, so Ron scooped up his daughter, along with the drum, and whisked her upstairs to bed. After a fairy tale and goodnight kisses from mom and dad, Sophie was in bed and her parents were alone at last.

They sat at the kitchen table, knowing the living room couch was not the right place for such a serious conversation. Caroline had ground some of the fresh coffee beans and the aroma was intoxicating.

Ron took his wife's hands in his. "So, tell me everything."

Caroline's mouth smiled but her eyes looked sad as she took a deep breath and began. She talked for an hour, only pausing to fill their cups with steaming coffee.

She started with the test results from Sophie, then to the Bone Marrow Registry, chemotherapy, Dr. Mansfield's brother-in-law who died, and finally about genetics and the second-child-option.

Just as he'd asked, she had told him everything. Then she dropped her head into her hands and exhaled. At first Ron thought she had begun to cry, but then she took her hands from her face and placed them on his.

"Ron, tell me something. Do you see us growing old together?" But before Ron could assure her of that fact, she continued, "I don't mean in a dream-like fashion of two people holding hands as they walk alongside a lake and talk about their grandchildren. Everyone imagines this. I mean, do you feel in your soul that we are not yet complete? Our time together? Because that's what I feel. I know it sounds crazy, but I believe our future has already been written for us, and that our life together is far from over."

"I don't think it sounds crazy. I think it sounds bold and courageous. We are going to overcome this together, and that will make US stronger, too. But, it's funny…"

"What?" Caroline asked her husband perplexed.

"Just what you said about our future having already been written for us."

Caroline raised an eyebrow. "How is that funny?"

"Well, it just reminds me of something that happened in Brazil."

"I'm glad you had such a good time." Caroline yawned. "I'm tired," she said.

"Me too. Let's go to bed then." Ron gathered up the coffee cups and placed them in the dishwasher. Then he took his wife in his arms and, after holding her for a moment close to his heart, he took her hand and led her upstairs.

Caroline's meeting with the oncologist, Dr. Blake, was not until 3 p.m., which gave Ron the entire morning and early afternoon in the office to get himself re-organized after being away. He wrote a trip report and presented his boss with the conference publications, which would be kept on file in the office for all the colleagues to review. These contained papers on what was presented at the conference, usually new design styles, as well as building methods and materials. The response about the Brasilia conference in the architectural community was already hovering on "ground-breaking".

Dr. Blake's office was located at South County Hospital,

which was where Caroline would also undergo the chemotherapy. If that was what they decided to do.

Ron and Caroline had both agreed to hear all the options that Dr. Blake had to offer and then they would discuss them at home before committing to one treatment method or another.

The waiting area at Dr. Blake's was large and sterile, with cold artificial lighting. The staff at reception was only accessible through a sliding glass window, which was closed when they approached. After waiting for a couple of minutes to be noticed, Ron finally tapped on the glass.

This was night-and-day different at Dr. Mansfield's office where the receptionist was always ready to greet them with a smile when they arrived.

A woman slid open the window. "I'll be with you in a moment." She then shut it again.

Caroline and Ron looked at each other in disbelief.

A moment later the window was opened again by the same woman. "Can I help you?"

Ron spoke. "Yes, we are Ron and Caroline Stanley. We have a 3 o'clock appointment with Dr. Blake."

The woman scanned her appointment list and checked off a name. She then handed them a clipboard and pen with several pages of medical history to be filled out, and the window was once again slid closed.

After 45 minutes of waiting on hard plastic chairs, they were finally shown into an office where they waited impatiently for another 20 minutes until Dr. Blake arrived.

"Good afternoon," he said, but they were not formally introduced. Doctor Blake sat down behind his desk and began reviewing the documents which they had filled out in the waiting room. Then he briefly looked through Caroline's medical file. "Unfortunately, we don't have a good donor match for you, Mrs...." Dr. Blake hesitated and looked back to his chart for Caroline's name.

"Stanley," Ron interjected before Dr. Blake could fill in the blank.

"Yes, Mrs. Stanley," Dr. Blake continued without apology. "From your charts, it appears that both your husband and daughter share similarities to your stem cells, but the chances of rejection are still quite high from both. Also, the donor bank did not have a match."

"But why do we actually need a donor?" Ron asked. "Can't chemotherapy or radiation alone treat the cancer?"

Dr. Blake took off his thin wire-rimmed glasses and placed them on the desk in front of him then took a deep breath. "In essence, treatment of cancer with chemotherapy is through the application of drugs that selectively destroy rapidly growing cells." The doctor spoke slowly and succinctly, so that they understood every word. "Extremely high doses, unfortunately, also kill the patient's own stem cells in the bone marrow and blood stream. For this reason chemotherapy should be done in conjunction with a bone marrow or stem cell transplant to prevent the patient from suffering grave effects from the loss of their stem cells."

"But how grave are the effects?" Ron asked. "I mean, doesn't the bone marrow regenerate itself? And if the chemotherapy kills the cancer cells—"

"It would be very ideal if it were that simple, Mr. Stanley, but bone marrow can't regenerate if there is nothing left of it.

The chemotherapy destroys it," he said in a somewhat condescending manner, as if his patience was being tested.

Ron understood that this was probably a monologue that Dr. Blake had given countless times in the past, to every new patient, but after the frosty greeting at reception and the long wait, it was all he could do to maintain his composure. For Caroline's sake he tried to stay calm. "And radiation treatment?" he asked in a respectful voice, hoping to avoid the same verbal scolding.

"Yes, radiation is also used to kill rapidly growing cancer cells. Unfortunately, high doses of radiation, especially in combination with chemotherapy, also kill the body's adult stem cells in the bone marrow and blood stream. This is why a bone marrow or cord blood stem cells transplant is performed, to replenish the patient with new, healthy cells." Dr. Blake began rubbing his left temple and looked down at some paperwork on his desk while he continued with the soliloquy. "In any event, radiation therapy would not be applied in this case. Radiation therapy is similar to surgery. It is a local treatment to eliminate visible tumors but is not typically useful in eradicating cancer cells that have already spread to other parts of the body." Straightening himself up as if impatient to move the discussion along, he added, "Therefore, I recommend that we begin with chemotherapy as soon as possible while simultaneously continuing to search for a donor."

Ron was aggravated and didn't want to give up. "And if we don't find a suitable donor? Shouldn't we at least try to use my cells, isn't that better than no chance at all?"

"Of course, Mr. Stanley. Of course. That is the path we would follow if a better match is not found." Dr. Blake handed a few brochures to Caroline and opened a copy of one of them to walk them through it. "This pamphlet describes the various chemotherapy cycles, and discusses the side effects in some detail."

"Will I lose my hair?" This was the most well-known of side effects to chemotherapy and a sure sign to the world of a cancer patient.

As though expecting this question, Dr. Blake continued. "Most likely you would lose a good majority of your hair, but it is not a permanent loss. It should begin to regrow immediately upon cessation of treatment. As I've said, this pamphlet discusses side effects in detail, which you can review at home. What we need to do now is to make sure you can maintain the treatment plan that I will suggest for you and schedule a series of appointments." His finger traced the printed notes on the page before him, but before he had a chance to continue Ron stood up and pushed his chair back.

"Doctor Blake, thank you for your time. We'll let you know what we decide in a day or two." By now Caroline was standing too, and holding her husband's hand.

For the first time, it appeared as though Dr. Blake realized that there were actually two living and breathing people before him, and not just theoretical patients. He seemed astounded by what Ron had just said. "What do you mean, you'll let me know what you agree on? From what I see in her charts, your wife needs treatment, and she needs it immediately. There are really no options here."

This time Caroline answered. "We understand, Doctor Blake. We'll be in touch."

They left the office and walked to their car without speaking. Once seated, Caroline slid over to her husband and laid her head on his shoulder. Ron slipped his fingers through her hair, then gripped her shoulder and held her tightly against him as she began to cry.

Chapter Twenty-One

"Hello, Mrs. Stanley," the receptionist said cheerfully. "I don't have you listed for an appointment today."

"No, not today, but I really need to see Doctor Mansfield. I'm willing to wait, of course." Caroline clutched her husband's hand as she spoke.

"Don't worry. I'm sure Doctor Mansfield will be able to squeeze you in. She does have a completely full schedule, but I'll let her know that you're here."

"Thanks," Caroline replied, then took a seat with Ron on one of the love seats in the comfortable waiting area. A coffee table in front of them was supplied with all the latest magazines, but neither had the desire to leaf through them at the moment.

Ron stood and walked to one corner of the room where he helped himself to a paper cup and filled it from the water dispenser. He drank it down, then refilled it and brought it to his wife.

"Thanks." Caroline drank the contents then crushed the cup in her hand. "That's what I'd like to do to Doctor Blake!" she said as she raised the demolished cup in the air.

Ron laughed, but couldn't help agreeing. "I suppose we

need to give him the benefit of the doubt; everyone has a difficult day now and again."

Caroline countered. "Yes, but of all people, that man should know from experience that for his patients, hearing about chemotherapy for the first time is one of the most difficult days in their lives. Talking to a doctor about intentionally putting drugs inside my body which will destroy living cells... I can't remember a much worse day, other than the day that I found out." She was clearly referring to the day when Dr. Mansfield first informed her about her cancer.

"I know, darling, I know. I understand, and I'm sorry if I was insensitive by sticking up for him."

"Oh, Ron, it's not you, don't apologize. I'm the one being too sensitive. I don't need an emotional caregiver, I need a qualified medical doctor to save my life. Maybe Doctor Blake is exactly what I do need."

Ron looked at his wife with skepticism.

She chuckled. "No, you're right. I don't need Doctor Blake. That's why we are here, to get professional advice from someone we trust."

As though perfectly timed, Dr. Mansfield walked into the waiting room and approached Ron and Caroline, with a smile in greeting.

Speaking in a low voice to keep their conversation relatively private from the others who were waiting, she asked, "Caroline, this is a surprise. I hope everything is alright?"

"Well, yes and no. I'm fine, nothing has changed. But I need to talk to you, well, about your gambling strategy, and I'd like Ron to hear your version of it."

Dr. Mansfield breathed a sigh of relief. "Okay, I'd be happy to talk to you both." She looked around the waiting room at the three other patients who were there. "It's going to be a bit of a wait. If you'd like, why don't you grab a coffee or something at the café next door, and come back in about an hour? There's no sense in sitting in here… doctors' offices are so sterile." She winked and Caroline and Ron both laughed.

"Good, we'll see you in an hour," Caroline confirmed.

The café next door was a New York-style delicatessen. Since it was late afternoon the sandwich crowd was gone and a few people were scattered at various tables drinking coffee. They ordered two lattés and found a small table on the side of the restaurant which overlooked a tree-filled park next door. For a long time neither said anything, they just sipped their coffees and watched two boys kicking a soccer ball back and forth in the grassy field with their father and mother watching nearby.

"Two kids, huh?"

Caroline looked at her husband. "Do you mean the two boys outside? Or us?"

"Us."

"Right."

They watched some more and then Caroline spoke. "You know, I would do anything to protect Sophie from harm."

"I know you would."

"And part of protecting her is protecting myself, because I know that she needs me to care for her." She began fumbling through her purse looking for a tissue.

Ron knew she wasn't finished so he sat patiently and waited for her to continue.

When she looked back up at Ron, her eyes were glossy

with tears and she dabbed them away. "The thought of her having to deal with my death is the biggest mental burden that I have at the moment. I hate that I can't control this. That it is in control of me."

"Honey, please don't think about—"

"No, let me finish." She took the last sip of her coffee, then added, "I don't think I can do it. I don't think I can bring another child into this world, knowing that it is highly likely that I would orphan that one too." After a deep breath, she added, "That child would most certainly grow up to despise me, knowing that I brought it into this world to try to save myself! That is so completely selfish."

"Please, Caroline, don't do this to yourself."

"But, Ron, don't you see? The child would also grow to resent you because you agreed to the absurd plan. I can't do that to you, and I can't do that to another child."

She seemed to have made up her mind as she was talking, right before Ron's eyes. She had so much to think about lately and he thought it was all too much for her to handle. He knew that he needed to try to keep an open mind, to keep all of the options on the table, but above all, he needed to keep himself well informed, so that together they could make the right decisions.

"Let's go see Doctor Mansfield," Ron said. "You like her, you trust her, and she can be objective. I think objectivity is what we need right now."

"I'm beyond objectivity, Ron. I think it would take a miracle now to change my mind."

Here's hoping for a miracle, Ron thought.

Reentering Dr. Mansfield's now empty waiting room an

hour later, the receptionist motioned for them to have a seat and said that the Doctor would be with them shortly.

It was already 5:45 p.m. Ron had called his mother to pick up Sophie at day care, so they could take their time.

Ten minutes later the last patient left and the receptionist led Ron and Caroline to the office where they had been together once before.

Dr. Mansfield was there waiting for them.

"Would you mind if I put on some music?" she asked to the mild surprise of Caroline and Ron. "It helps me relax, and it looks as though you two may need it more than me right now."

Ron laughed. "That's true. Please, go ahead," he said as he motioned to the stereo.

A moment later the light-as-air strings of Mozart danced through the room, helping to put all three just a bit at ease.

"So, how can I help?"

"First of all," Caroline began, "thanks for meeting with us on such short notice." The doctor waved her hand in a gesture implying there was no need for thanks and Caroline continued. "I'm actually not really sure if we need to discuss this further, but I promised Ron that we would talk to you together about it."

"I'm assuming you are talking about the second-child option? As we somewhat apathetically call it," Dr. Mansfield remarked.

"Exactly. Apathetic. Selfish. Cruel." Caroline stated.

Ron took his turn, trying not to come across as either argumentative or close-minded. "Now wait a minute, Caroline, let's seriously talk about this. You may be right, but you also may be wrong. Let's at least discuss this and come to a mutual agreement." Ron's voice was sensitive but strong, which was apparent when he saw the ascending nod from his wife. He then turned to the doctor. "Doctor Mansfield, could you please, for my benefit, explain this theory."

"Of course, I'd be happy to. But can I first ask, have you spoken with Doctor Blake yet?"

"Yes, which is actually why we are here."

"I don't understand," Dr. Mansfield answered.

Ron continued, "We both met with Doctor Blake today. And... well... I'm not sure how to put it politely but let's just say there was no chemistry between us."

"Ah, I see. This isn't the first time I've gotten that kind of feedback about him. But," turning to Caroline, she added, "I've told you before, I'm not an oncologist. I can't oversee your chemotherapy treatment."

"I know. And the truth is, I'm not sure that I want to begin chemotherapy. I just wish there was another choice."

"Caroline, we unfortunately don't have the luxury of options, and from your previous comments, it sounds as though trying to conceive a second child is not on the table for you. Please don't tell me you are giving up."

Caroline wasn't sure how to respond. She simply looked down at her hands in her lap and waited for someone else to continue the conversation. It was Ron.

"Doctor Mansfield, we are not giving up. We are going to overcome this. We just need to get our battle plan drawn before going into war." This garnered a smile from his wife. "Could you please tell me how this would work? I mean, can she undergo chemotherapy treatment while trying to get pregnant? Or while pregnant? Because, if she were to conceive and the baby is a match, the transplant would have to take place as soon as possible, correct? And the chemotherapy would need to take place prior to that time, to kill all the cancer cells. Am I right?"

Doctor Mansfield took her time and explained everything, from how in vitro fertilization and implantation could increase the chances of a quick pregnancy, the statistics of the perfect-match-baby, the possibility of sterility from the chemotherapy, to the timing of chemo and transplantation, if all goes well.

"So, it's a gamble," Ron concluded. "And essentially we have to make one of two choices right away."

Doctor Mansfield nodded, allowing him to continue.

"Either she begins chemotherapy immediately and hopes that a matching donor is found..."

"Right..." the doctor replied.

"Or she forgoes chemo in an attempt to conceive a child which would be an exact match."

"That's correct."

He took his wife's hand in his. "Then, once the child is born, she would have to begin chemotherapy immediately."

"Yes. Or even earlier than full term. Once the baby is viable it could be delivered via Caesarean section," Dr. Mansfield appended.

"And then, once the chemo is complete, she would go ahead with the transplant."

"Exactly."

Then Caroline added, "Or if the baby is not a match, hope that another donor has been found and that it is not too late."

"That pretty much sums it up," Dr. Mansfield confirmed.

There was silence in the room and both Ron and Dr. Mansfield had their eyes on Caroline.

"What do you think, honey? Should we try it?" Ron asked. "It could work."

Caroline shook her head, her voice was almost a whisper, and a tear began to slowly roll down one cheek. "I'm sorry, Ron. I can't do it. I can't risk another baby. I'll have to start with the chemotherapy. I just can't do it."

Later that evening Ron sat at his computer once again, the GIFs on the Google homepage dancing before him. But he wasn't sure what to search for—information on chemotherapy, survival stories or perhaps other oncologists in the area since it wasn't clear if Caroline would be going back to Dr. Blake.

The two hadn't spoken much that evening. Both were exhausted, both physically, due to Caroline's illness and Ron's travelling, and emotionally, from the doctor's appointments.

Caroline had gone to bed an hour previously but Ron couldn't settle down. He needed to take action. He just wasn't sure what to do.

Why couldn't this be more like designing a building, where all the numbers have to add up? Why couldn't he simply line up the facts and make an intelligent decision? If he only knew what the future would hold, if only he could have a brief glimpse into the future, to know what only God knows, to know if his wife would still be here ten years from now.

And then it hit him. The answer was so clear. How could he not have thought of it before?

Then he buried his face in his hands and thanked God.

Chapter Twenty-Two

"You've been offered a professorship? Ramon that's wonderful news!" The happiness in Maria's eyes in response to his announcement filled Ramon with pride.

He had been invited to dinner with Rosa and Maria after having spoken to Rosa a couple of times on the phone since their first meeting.

Rosalinda also voiced her support on his news. "Congratulations, Ramon. Eduardo would have been very happy to know that you have chosen to teach. And I'm glad that the University was smart enough to know that they should keep you and your talents here. From what Eduardo told me, you have more than earned the honor of that position."

"You are too kind. Certainly, it is what I wanted, to teach that is, and to remain here in Sao Paulo. Of course I need to complete my dissertation, but I can finish that simultaneously since they need me to begin teaching right away."

"But will you be teaching all of the courses from Papa and Professor Montagne? That would be a lot to start with," Maria asked with a concerned look.

"No, I'll start with your father's courses, and Professor Montagne's courses will be redistributed amongst the remainder of the teaching staff. Then Professor Kumar will evaluate whether an additional instructor should be hired." Turning to Maria to clarify, "Professor Kumar is the Head of the Math Department, and the one who offered the position to me."

"Oh, yes, I remember him from the funeral. He seemed like a nice man."

"Yes, he *is* fine to work with, though he's strictly involved in running the department and not in conducting research. But now he has also had to take over some teaching."

When they finished eating and the dishes had been cleaered, Rosa had a proposal. "Maria, why don't you take Ramon out into the garden. I'll prepare some coffee and join you in a moment so we can discuss your father's funny equation again."

"Oh, I'll help you with the coffee mother."

"No, Maria," Rosa squeezed her daughter's hand slightly and gave her a gentle smile, "stay with Ramon."

And she bustled off into the kitchen leaving her daughter momentarily caught off guard at being alone with their guest.

Twenty minutes later with a tray in hand, Rosa stepped out onto the terrace and began arranging coffee cups on the small table. Her daughter and Ramon were on the far side of the garden laughing and didn't hear her come out. She sat down and watched the pair with contentment until they finally noticed that she was there. They quickly joined her.

After the coffee, cream and sugar had been doled out, Rosa had a curious question, "Tell me Ramon, does Professor Kumar know about Eduardo's lifespan equation?"

Ramon paused for a moment to think before answering. "I'm not sure. In any case, he has said nothing to me." Then he asked, "Does this concern you?"

Taking a deep breath and nodding slightly, she continued, "Yes, Ramon, I've been thinking quite a lot about the equation and I do think we need to be very careful with this information. Eduardo made it clear that it is a very powerful tool. We all saw this first-hand with the example of Professor Montagne." She took a sip of coffee and then gently replaced her cup on the delicate saucer. "But I would like to discuss this openly with you, Maria and Miguel before making any decisions about whether or not to destroy it."

"You are right, Senhora."

"Please call me Rosa, Ramon."

With a cordial nod, Ramon continued, "You are right to have concerns. I've also had this heavy on my mind. This equation is revolutionary. Once the information is out, it will travel quickly around the world. Everyone will want to have it, and your family would be right in the middle of the attention. It would drastically change your lives."

As Rosa looked around the quiet garden and felt the comfort of her present company, she knew with certainty that what she had right now was all that she could ever want.

Professor Amar Kumar was walking back to his office after a three-hour-long meeting with the university regents. His attendance at such meetings was one of his job requirements that he was not particularly fond of. Topics typically ranged from fund raising, student solicitation and new construction, to faculty nominations.

The nomination of Ramon Santos to the Mathematics Faculty, for which prior approval had been confirmed at his request, was briefly mentioned but Kumar was never given time to speak nor was his opinion ever asked. This appointment essentially got his department out of the difficult spot of being dreadfully under-staffed after two of his teaching staff died within days of each other!

As usual, the Regent meetings always left him with a feeling of impotency, whereas his presence in the Mathematics building gave him the security and sense of superiority which he so craved. He hurried his pace to return to the refuge of its four walls.

Since the walkway to the front entrance was being repaved, Kumar entered through the rear door. As he turned down the hallway to his office he could hear his secretary's voice. Her tone was elevated and he could immediately sense her frustration. There were short pauses in which he heard no second voice, so he realized she was on the telephone. Just as he entered the office, she slammed down the phone.

Kumar stopped in his tracks to find out what had gotten his otherwise complacent secretary so keyed up.

Startled by his presence she immediately began to apologize for her behavior. "Oh! Professor Kumar, I didn't hear you come in. I'm sorry about that..." nodding towards the phone, "but it just won't stop!"

Still unable to slow herself down, Kumar tried to find out what the problem was. "Please calm down, tell me what the problem is."

"It's crazy! The phone has been ringing non-stop since you left." As if on cue, it began to ring. "See?!?"

"See what? Who keeps calling?"

But she didn't answer his question, instead she ran over to the window. "And look!" She was now pointing to the front of the building as Kumar approached her side.

Peering out the window he saw what appeared to be a group of about 20 people gathered. His first thought was simply that it was several students talking, and he was just about to turn away and dismiss his secretary's nonsense when he noticed that the group outside was not equipped with school bags but with cameras and microphones.

They were not students. It was the media.

"What's going on?" he now demanded of his secretary.

"I'm not completely sure, but I assume it has to do with the phone calls."

He now looked sternly into her eyes, "Ana Maria, tell me what the phone calls are all about."

She collapsed into her chair and began to talk more calmly. "It's the press. They want information about some work that Professor Rodriguez was involved with."

"Eduardo?"

"Yes. I keep trying to tell them that he passed away and they just laugh! Can you imagine? They seem to think his death is a big joke. They were even asking me when I found out that Professors Rodriguez and Montagne were doomed!"

Kumar tried to think. "Okay, Ana Maria, please try to stay calm, there must be a misunderstanding here. For the time being, do not answer the telephone and do not speak to any of the reporters."

He headed for the door. "I'll find out what is going on here."

Professor Kumar came out the front entrance and onto the steps. For the first time he was actually happy that the front walkway was undergoing reconstruction, as it gave him a buffer zone from the reporters. They immediately noticed his presence and gathered as close as they could.

One of them shouted, "Are you Professor Kumar?"

"Yes, I am. And may I ask who all of you are and why you are here?" The brief feeling of control he felt over the reporters was quickly washed away by the following barrage of questions.

"Do you plan to use the equation on yourself, Professor?"

"Has it been patented?"

"Does the university own the property rights?"

Kumar felt his pulse rise. He had no idea what they were talking about. He was about to turn around to go inside, but wondered who he could call to find out what was going on? Maybe he could find out something from them first. He faced the reporters and tried another tactic—pretending to know exactly what was going on. "Who gave you this information?" he asked. He was given no response to his question but overheard some rumblings about the police department.

The police department? Kumar wondered what the math department had to do with the police. "Tell me exactly what you've heard and I'll confirm or deny it, but that's all."

One of them called out, "That Professor that died..."

He was interrupted by laughter and several in the group were heard asking 'which one'? That amused the assembly even more.

The reporter continued, "Yes, the first one, Rodriguez. We've heard that he developed an equation that predicts your own death. Apparently, you can just sit down at the computer and figure out when your time is up."

Another reporter spoke up. "Is it true that both Rodriguez

and Montagne used that equation, and now both of them are dead because of it?"

Kumar' mind was racing, what were they talking about? He needed to find out the source of this information, so he asked, "You heard this from the police?"

The reporter spoke up again. "So, are you confirming or denying it?"

Another called out, "Well, he didn't deny it, so it must be true."

"I've confirmed nothing," Kumar stated. Not only did he have no idea what they were talking about, but he wasn't sure why they found it so amusing.

Kumar's frustration was now running high. He wanted answers, and he wanted them immediately.

He turned and stormed back into the building. Stopping at his secretary's desk he asked, "Ana Maria, could you please call the police department and get me on the phone with the officer investigating Professor Rodriguez' death?"

She immediately picked up the phone.

"Connect to my line as soon as you've found him." Kumar hurried into his office and shut the door behind him.

Chapter Twenty-Three

As a little girl standing on the ground looking at an airplane flying thousands of feet above, Sophie had always wondered if the people in them were afraid of falling out.

But now, sitting in one herself, she wasn't scared. She peered through the window at everything below. Nothing looked real. It looked like a toy village with tiny houses, winding roads, forests, mountains and farmlands. Eventually she saw nothing but the blue of the ocean water, as far as the eye could see, and then suddenly nothing but a bed of clouds. Her mind was full of thoughts and questions and, even if she'd wanted to, she couldn't take her eyes away from the airplane window.

Caroline held her husband's hand on her left, and her daughter's on her right as they came into their approach to Brasilia. The sun was low in the sky and the view of the city was spectacular. She almost forgot for a minute why they were there. She wanted to forget, so she closed her eyes and dreamed of flying to a sunny vacation on the shores of a remote Pacific island.

"Are you alright?"

Caroline was startled back to reality by Ron's voice and opened her eyes just in time to feel the first bump of the plane on Brazilian soil. She laughed. "Yes, I'm fine." Another bump, and then the plane leveled itself. "Just daydreaming." She closed her eyes again and smiled when she felt Ron's lips gently brush her cheek.

"We're here!" Sophie unlatched her seat belt and tried to squirm past her parents.

"Hold on, Sophie, the plane hasn't stopped yet. Same rules as in the car."

"But—"

"No, 'buts'." Caroline lifted her daughter back into her seat and refastened the seat belt. "And no taking off your seat belt until the captain says so. Okay?"

A frustrated sigh was her only response.

Miguel was waiting behind the gates once they passed through Customs. He and Ron shared a warm handshake.

"Thanks for meeting us, Miguel. It's good to see you."

"Good to see you too." Then turning to Caroline and Sophie, he added, "And who are these two beautiful women that you have brought with you this trip?"

"Hello, I'm Caroline." She never liked to be formally introduced and preferred to offer her name herself.

"A pleasure to meet you, Caroline." Miguel responded as he handed her one of the two yellow roses in his hands.

"Is the other one for me?!?" Sophie jumped up and down on her toes in earnest waiting for Miguel to hand her the flower.

"Well, that depends," he answered.

"On what?" she countered.

"Well, if you're name is Sophie, and that's your mom holding the other yellow rose, then this one belongs to you."

Miguel answered with a grin while pointing to the beautiful flower in his hand.

"It's me! It's me!"

Miguel couldn't hold back any longer and handed her the flower, to Sophie's delight.

They followed the flow of foot traffic towards Baggage Claim, but along the way Sophie suddenly stopped and tugged on her mother's hand, "I need to pee-pee!"

Caroline looked around and spotted a ladies room, which she pointed out to her husband. "I'll go in with Sophie. Should I just meet you at Baggage Claim?"

"No," Ron said, "we'll wait right here. Go ahead."

Once the girls were gone, Miguel said what was on his mind. "I'm glad you decided to come back, and show your family our beautiful country. Although, I have to admit that I'm very surprised that you've decided to come back again so quickly." Ron was about to respond but Miguel continued, "You mentioned on the phone that it was because of Caroline's illness that you are here, but quite honestly, she looks to be the picture of good health. Is it that much worse than it appears?"

Ron was cautious about his answer. Could Miguel possibly know why he was there? He did not want to delve completely into the topic just yet. He wanted to wait and do it when they were sitting comfortably and Caroline was there with him. So he simply answered the question, "Yes and no. She is doing well at the moment, but if she doesn't get treatment soon that will change very quickly."

Miguel nodded in an honest sympathetic gesture, "Oh, I'm really sorry, Ron. I guess that is the reason for the urgency."

The girls emerged from the ladies room, but before they were in earshot Ron added, "Actually Miguel, Caroline and I would like to talk to you tonight, with Katie, and then you'll know exactly why we are here."

Ron had booked a room in the same hotel where the conference was held the week before. Caroline and Sophie went up to the room while Ron and Miguel headed to the service desk to pick up the reserved rental car.

"I really don't mind driving you back here after supper, Ron."

"Thanks, Miguel, but you've been such a big help already by picking us up at the airport and having us over. Plus, having our own car gives us some flexibility to see the sights. Caroline can't do too much walking, especially not in the hot sun."

They followed Miguel to his house and parked the car on the street. Just as it happened during the first visit, they were greeted by three little kids in the driveway and Katie coming out the door. Sophie was immediately surrounded by children and skipped off towards the house holding Rosie's hand and singing, "Bye, mama, bye, daddeeeee!!!"

Ron was now holding his wife's hand and they walked up the driveway together. He gave her hand a brief squeeze before he released it so that she could greet Miguel's wife.

"Hi, I'm Caroline," she said extending her hand.

Katie didn't seem to notice her outstretched hand and instead placed both of hers on Caroline's shoulders and gave her a kiss on each cheek. "Very nice you here, Caroline. I am Katie."

Caroline felt immediately comforted at the warm welcome. Ron then greeted his hostess as if she were an old friend, after which all four followed the kids into the house.

This time the table was set for eight and the kids joined them for supper. The language barrier for Sophie and her new friends was no obstacle at all. They each spoke their own mother tongue and filled in the blanks with hand gestures and laughter. The meal was lively and relaxed, as if they'd all been friends forever.

Once all their bellies were full, Caroline helped Katie with the dishes while Miguel and Ron brought the coffee out to the dining table. Meanwhile, Sophie was being well entertained with dolls in Rosie's room, while the two boys were busy with a computer game.

The four adults returned to the table and there was suddenly an uneasy silence, as if all of them knew what the next topic would be, but no one wanted to begin. Finally, once the coffee had been served, Ron began his semi-prepared speech.

"Miguel, Katie, I know you both must be wondering why I have come back so suddenly to Brasilia with my wife and daughter. I thank you for inviting us into your home and we both hope that one day we will be able to welcome you to ours."

All of a sudden it was difficult to approach the topic and he began to wonder if these people who had quickly become his friends could actually help him. And whether or not they would, even if they could. Would they turn their backs on him and his family? After all, he hardly knew them, only a couple of weeks ago they were strangers.

Miguel tried to help. "Ron, go on. We're listening."

That small encouragement was all he needed to let the

flood waters run out. He talked about Caroline's specific illness, her doctors, her prognosis, the insurmountable odds of them sharing HLA molecule types and the disappointment of no suitable donors in the family or in the registry.

Translations were needed several times for Katie, who kept giving positive words of encouragement in between.

"Thanks for your kind words," Caroline said, "but we are not here for sympathy, in fact, we have a request."

"Sure, Caroline," Miguel said, "we'd like to help if we can. Please ask."

Caroline turned to her husband who then continued for her.

"Thank you, Miguel," Ron began, "so, this is what it comes down to. As I have just described, we have a one in four chance that a child of ours would be a perfect donor match for Caroline. Sophie is not a match. Statistically our chances of a second child being a match are one in four, but Caroline is afraid that she will not live long enough to bear a second child, or if it is not a match, that she would then leave two children without a mother." Ron could see a tear roll down his wife's cheek but he continued. "She won't do it." He shook his head. "She won't do it unless she knows that the child will be a match." He then stopped talking for a moment to be sure that Miguel was following him. Katie nodded in politeness but it was clear that Miguel would have to fill in the blanks for her later.

Miguel then asked, "But, is there any way to know if the child would be a match? That's impossible, isn't it? Or are you talking about testing it while it is still a fetus?"

"No, I'm talking about knowing even before she became pregnant."

Miguel immediately shook his head, "But, that's impossible."

"That's exactly what I thought too, until I remembered a part of our conversation, right here, last week."

"Here? I'm sorry, Ron, but you've lost me."

He hesitated for just a moment and took a sip of coffee. "There is a way. And it involves what you told me and Joe after you got off the phone with your sister last week. It's your father's work—"

"Ahhhhhh," replied Miguel, "now I understand."

Ron was relieved at how quickly Miguel seemed to follow his reasoning.

Miguel turned to his wife, "Equação do pai."

Katie had a surprised expression on her face.

"Seems like the whole world wants to get their hands on this thing," was Miguel's next comment although it was unclear to whom it was directed. He seemed lost in thought.

Ron thought the air in the room suddenly seemed stifling and Caroline must have thought so too as she excused herself to get some fresh air. Ron was worried that he had suddenly lost his last chance and tried to back-track.

"Miguel, if we know how long she will live, then we will know if the baby is a match. It's that simple. And if your father's equation can give us that information, then it may save my wife's life."

"Please," Miguel said, "I'm not upset by your request. My reaction is simply to the mention of my father's work."

Miguel then stood up and Ron, disheartened by what he

perceived an end to their conversation, also rose from the table.

Miguel quickly turned to him and smiled. "Please sit, my friend." Once Ron was seated, Miguel clarified things. "There has been a lot of attention being paid to my father's work these past few days, most of which makes me wish we'd never discovered this secret equation of my father's. But, to tell you the truth, your interest is the first positive mention of it that I've heard at all."

"It's such a relief to hear you say that, Miguel."

"Why don't you get Caroline back in here while I get some glasses and the cognac." Miguel smiled. "We'll have a drink while we talk this through."

Chapter Twenty-Four

"But of course the proprietary rights belong to the university." Professor Amar Kumar was putting forth his most confident voice in the hopes of convincing his caller. "Our professors are obliged to sign a contract foregoing rights to research conducted at our facilities."

"In any case, the university has not yet decided as to whether the equation will be kept for further development here, or if it will be auctioned off. If you would like to leave your name and telephone number with my receptionist, I would be glad to contact you if a sale is imminent." And with a click his caller was transferred back to Ana Maria, who'd been taking names and telephone numbers all day.

Amar was frustrated. He had been able to contact Inspector da Silva the previous evening and was given the details of the discovery of the equation and the involvement of Professor Montagne, but despite searching both offices of the deceased professors, nothing even remotely resembling such an equation was found.

Unfortunately, the only copies were out of his hands, and although he was not by any means sure whether the university, or more specifically, his own department, had proprietary ownership, he was ready to begin an all-out battle to claim it.

To make matters worse, a call to the Director of the University Regents to discuss the topic was not returned, although the Director's receptionist assured Amar that the Director was aware of the issue. Aware of the issue! Amar could not understand how it was possible that the University Directors had not contacted him about an issue directly related to his department.

Then the telephone rang again to wake him from his reverie. He hoped it would be the Regents Director, and waited anxiously as he let his secretary take the call.

Ana Maria answered. Then, after several moments, she transferred the call to Amar's office. "I think you'll be interested in this one, Professor", she called through the open door.

Perhaps it was the call he'd been waiting for. "Professor Kumar speaking."

"Hello, Professor. My name is Jonathan Boyd."

Or perhaps not.

The caller continued, "I represent WP Pharmaceuticals in the United States. I'm sure you've heard of us."

"Yes, of course I have, who hasn't?" It was only the largest pharmaceutical company in the world, known for its cutting-edge biotechnology research and development programs and enormous energy focused on controversial issues such as stem cell studies and cloning.

"Right, well, my company will be taking this new equation of yours off your hands—"

Amar cut him off. "I'm sorry Mr. Boyd, the equation is not yet for sale."

"I'm aware of that, Professor, that is why I am calling to give you notice. I expect that by tomorrow you will have it available 'for sale' as you put it. I will be at your office at noon, at which point I will collect all electronic media and any printed or hand-written notes related to the equation. ALL copies, that is."

Amar couldn't hold back his laugh. "Please, don't bother, I haven't yet decided—"

Now it was Mr. Boyd who interrupted. "This is not a matter for you to decide, Professor. I've spoken with the Regents. The deal is done. Over the next few days we will either validate the work or refute it. At the end of that time, an eight figure sum will be transferred to the university if this equation is genuine. If it does not perform according to its purported function, it will be duly returned to you."

"But—"

"Good day, Professor. I'll see you tomorrow."

And the line went dead.

Chapter Twenty-Five

Rosalinda was busy in the kitchen brewing fresh coffee when the doorbell rang.

"Oh, Professor Kumar! This is a surprise. Please come in. I was just making some coffee."

"That sounds wonderful, Senhora, if it isn't too much trouble." He hadn't seen her since the funeral and was wondering whether or not to give his condolences again. But her generous and open nature made the idea seem unnecessary.

Actually, the only time that he had been to the Rodriguez family home had been the day of the funeral. He had known Eduardo for many years, but they had never had much more than a polite professional relationship. Eduardo had always treated him with respect, even when Amar was only a professor and not yet Department Head. But for some reason, Amar felt that the deferential handling that he received from Eduardo had nothing to do with his high position, but rather that Eduardo treated everyone respectfully the same, simply because that was his nature.

His wife seemed to be of the same mold. She was polite, gracious and cordial.

She emerged from the kitchen holding the coffee tray and

found Amar was still standing in the entrance way where she had left him.

"Please, please, sit down, senhor", she said, as she ushered him into the living room.

He felt ridiculous after having been daydreaming in the entrance and now began reviewing in his mind the strategy that he had come up with overnight to recover whatever he could from Eduardo's home files. He anticipated his mission to be an easy one, with Senhora Rodriguez in mourning for her husband, not to mention lonely and most likely wanting to be helpful in any way possible. But before he could speak, she drew a serious look on her face after handing him a cup of coffee and then came right to the point.

"Professor, I can say that your visit actually isn't a complete surprise to me."

"Is that so?"

She raised an eyebrow and continued, "Yes, it is so, Professor. And quite honestly, I hope you'll be content leaving here with no more than a belly full of coffee."

At this remark Amar almost spit out the sip that just entered his mouth. "Excuse me? What do you mean?"

"Please Professor, I may not be university educated, but I do watch the news, and what you want, what the whole world will soon want after last night's headlines, I cannot give to you."

"You can't or you won't?"

"Both."

"But you do possess what I seek?"

She drew a deep breath and leaned back into her chair. "Professor, I'm not really sure how you are caught up in all this, but I highly doubt that you have anything personal to gain, other than an attempt at pleasing the Regents. But you know as well as I that they are just using you to get what they want."

Amar opened his mouth to speak, but his hostess raised her hand to signal that she wasn't through.

"My husband was very clever, but I'm sure you already knew that. Not only was he a brilliant mathematician, but he was clever in terms of human nature too. He had a feel for people and procedures and knew how to deal with them both. He always came out the good guy, which he was of course."

"I'm sorry, you've lost me."

"Let me also apologize, professor Kumar. I don't meant to be rude. I am very tired after all that has been going on recently." She took a deep breath before continuing. "You are a smart man, and my husband had a high degree of respect for you. And despite my forthrightness, I know that you understand me all too well. Eduardo has outsmarted all of you who are interested in his work, because none of you will get it."

At this point Amar realized that he was not going to make any progress by being kind and appearing to be doing the right thing, so he decided to change tactics. "Senhora, you must realize, all of your husband's work was sponsored by the university. Anything of scientific value belongs to that institution, and not to you or any other private entity. Whether you provide me with the information I seek now, or give it to the police later, is simply a matter of time and logistics."

"I disagree."

"Whether you agree or not, the law is the law, and I'm afraid that eventually you will have to supply me with what I am here for. Wouldn't it just be easier to hand it to me now and avoid any future complications?"

Rosalinda stood up as an indication that it was time for her visitor to leave. "Quite frankly, I'd rather destroy the work than pass it on to you or anyone else."

"Then I'm afraid that you would have a different kind of trouble on your hands, Senhora, one that would likely involve criminal punishment on your behalf."

Rosa opened the front door for her guest and left him with one last remark. "I'm an old woman, Professor, and one of the many things I've learned in my life is that doing the right thing should never cause one to fear." And with her guest out on the front step, she closed the door firmly behind him.

After work that evening, Maria stopped by her mother's house, as she so often did, but the simple chats that they used to have about daily life were no more. Rosa filled her daughter in about the events of the day, including the visit from Professor Kumar. Maria likewise told her mother what she had heard from her colleagues in the office.

Everyone was talking about the equation, as if it was a revelation, a sign from God, perhaps the Second Coming. But at the same time it seemed to empower the science community. What would they discover next? Our predestined mates? The number of children we will have? The meaning of life?

The equation may be the mathematical solution to the spiritual theory of predestination. The first step in solving God's mysteries. Theologians and philosophers would be interested in using it, each applying it to their own theories. There were applications too numerous to count.

There was a knock on the front door and Maria got up to

open it while her mother warned her to be cautious. But she knew that already. She was as weary as her mother from the constant media attention and answering, or rather avoiding, their questions.

But when she opened the door this time, and called her mother to her side, both knew immediately that the situation had risen to a whole new level.

Three men stood on the front steps. All were dressed the same, in long purple robes, matching skull caps, and large golden crosses on chains hung around their collars. Each stood with their hands crossed before them.

"Good evening my child," the man in the middle addressed Maria with tenderness, then turned to Rosa, "madam."

"My name is Cardinal Pena, this is Cardinal Rathborne," he gestured to his left, "and Cardinal Menendez," motioning to his right. Turning his head slightly to nod towards the presence of the men behind him, he added, "and these are our guard dogs." There were four men dressed in dark suits and dark glasses who were focused intently on the Cardinals, the two women, the premises and the passing traffic, respectively. The cardinal then smiled as if to say the guard dog comment was a joke.

Neither Maria nor Rosa had yet said a word when the man continued, "His Holiness has sent us with a message, may we please come in?"

"His *Holiness*? As in His Holiness *The Pope*?" Rosa asked.

"That is correct, Senhora. We would very much like to speak with you... in private."

Rosa noticed the neighbors across the street peering out of their front door, and she motioned for Maria to let the visitors into the house. Two of the security guards remained outside and the other two followed them in. Rosa ushered them into the living room and offered them to sit.

"Would you like some coffee, Cardinals?"

"No, thank you, Senhora Rodriguez." Cardinal Pena answered for himself and the rest of them. He then sighed deeply. "Senhora, I'm sure you know the reason that we are here."

"Yes, I'm sure I do," she responded in agreement.

Cardinal Pena brought his hands together so that just his fingertips were touching, then he asked, "Are you a believer, Senhora?"

Although impressed by her visitors, she was still hesitant to answer almost any question after her experience with the press. "A believer of what, Cardinal? God or my husband's equation?"

A look passed among the Cardinals.

"Senhora, please excuse our appearance, that is, our security and our extravagant robes, but we were sent here today by His Holiness to work in the name of God, for His sake. And it is for His sake, and for Him, that I ask if you are a believer."

Rosa caught her breath at his humble nature and realized that she may have misjudged them. "I apologize for my abruptness, but as you can imagine, you are not the first to inquire after my husband's work."

She hoped that this would appease them, but to be sure she then added, "Yes, I am a believer, and so was my husband. God rest his soul."

"Very well, Senhora Rodriguez, then I'm sure you will understand our concern here."

Rosa was not sure about anything at that point, although she did have a gut feeling that they were on her side. "Until now, everyone has been concerned about the money they could make from my husband's work, but I think your intentions must lay elsewhere, am I correct?"

The Cardinal gave her a warm smile, "Yes, our interests lie elsewhere, and that is, in the saving of souls."

"But how can this equation of my father's save souls?" Maria could no longer stay quiet.

The Cardinal responded with directness to Maria. "That is exactly the problem. This equation would do just the opposite."

Rosa gasped at the thought. She did not understand how this man could say such a thing. She would never believe that her husband's work could in some way hinder the saving of Christian souls!

But before she could speak the Cardinal continued, "Please let me explain." He took a deep breath then closed his eyes.

For a moment his lips moved ever so slightly, then he re-opened his eyes and they looked onto Rosa with what appeared to be a perfect peace. "Senhora, faith is something that stems from our hearts and extends to our souls, and in the process, our entire bodies become consumed by it. Our belief and love for God is unquestionable and many believe it is unexplained. Which is the problem with your husband's work. A 'scientific explanation', if you will, could, for many people, replace a need for faith."

"But that is not so!" Maria interjected. "My father was a strong believer in God, he would never have had such an intent for his work!"

The Cardinal was still as passive and calm as before Maria's outburst. "My dear, I am not in any way accusing your father of heresy. From what I understand, he was a good man, and an excellent mathematician. His integrity is not in doubt."

But Maria was still not pacified. "Then, why are you here if not to discredit him?"

The serious look on the Cardinal's face stopped Maria's inquiry and she listened in earnest to his response.

"We are not here to discredit anyone, my child. Quite the opposite. We are here to save the reputation of our beloved Church and, with that, the gift of our own salvation."

Chapter Twenty-Six

It was almost easy for Caroline to forget why she was there, in Brazil, on what should have been an adventure, the trip of a lifetime.

After meeting with Miguel on their first night, they toured the city the next morning and she marveled at the differences from home. Everything was new and foreign—the language, the way the people looked and dressed, the food, the architecture, the trees and plants—all except for the McDonald's restaurants. Sophie wanted to eat there of course, but both Ron and Caroline were adamantly opposed. While in Brazil they wanted to expose her to as much foreign culture as possible, including the food, and McDonald's was not typical Brazilian fare.

And now, as though time was speeding by, they were sitting on a flight to Sao Paulo along with Miguel, on their way to talk personally with his mother and sister about the use of the equation. Not only were they eager to talk about the equation, but Rosa had called Miguel that morning, pleading with him to come as soon as possible, telling him how things were getting out of control and that she had to make a final decision about the equation right away. However when he confirmed that he would come, he did not mention that he was bringing along a couple of friends from the United States.

Miguel had explained to Ron and Caroline that the ultimate decision concerning his father's lifespan equation belonged to his mother, to whom his father had left the rights. But in the short time that Ron and Caroline had been away from home, they had not had time to see or hear any news and were oblivious to the storm that was brewing over the equation. They thought they had privileged knowledge. Miguel dashed those hopes when he further explained that his mother had been hounded by the press since the equation became public, and that she'd already hinted that she wanted to have it destroyed.

They'd have to convince her to hold off on that decision for the time being if they were going to be able to use it on Caroline.

The flight was only an hour and a half and went by so quickly compared to the recent twelve-hour haul out of New York. But despite the heavy travelling, not to mention Caroline's illness, none of them felt tired. They were running on adrenalin and anticipation. Every moment brought something new.

As they exited the airport gate, they were greeted by a beautiful young woman who warmly embraced Miguel.

Before they could ask whether he had a girlfriend in Sao Paulo, he introduced her as his younger sister Maria. She was quite surprised too, as she was only expecting her brother. But they were still able to pack themselves and their luggage into her car, with Sophie, Ron and Caroline tucked cozily in the back, and they were soon darting through the streets of yet another fabulous foreign city.

Sao Paulo was very different from Brasilia, and although they weren't travelling through the typical tourist areas of the city, Miguel pointed out anything and everything of interest. This was where he was born and raised, he knew every turn, and was clearly passionate about the city, showing an emotional connection which had been lacking during the tour of his new home city of Brasilia.

Maria's knowledge of the English language was limited and since she seemed to have a lot on her mind, the conversation in the car was primarily between her and Miguel. But Caroline didn't mind, she enjoyed listening to Maria's sweet voice bubbling on and on with her brother, clearly excited to be with him. And Caroline was just as happy to keep her attention focused on the unfamiliar sights outside the windows.

Stopped at a traffic light, a large black SUV with tinted windows pulled up beside them. Maria didn't seem to notice, as she continued with her stories, but Miguel couldn't help but be curious, since the vehicle was sorely out of place.

Sophie was unusually quiet, as the new sights were extremely foreign to her and a bit overwhelming, so she held tight to Caroline's hand.

Miguel pointed to a gas station up ahead and, although Maria initially shook her head, it appeared that Miguel insisted she pull in. She didn't stop the car at the gasoline pump, but rather in one of the parking spaces, and both she and Miguel got out of the car.

"What's going on, daddy?" Sophie asked.

"I'm not sure Sophie, maybe they need to use the bathroom." But instead of going into the building, the two simply circled around the car and switched positions, with Miguel now taking over the driving.

"Is it much further?" Ron asked.

Miguel looked into the rear-view mirror and hastily pulled back into traffic. "No, not far, but we are going to take a bit of a short cut." At that, he pulled quickly onto a side street and with a series of aggressive turns and bursts of acceleration, they were deep into a rather poor housing area.

As Ron hadn't seen this type of driving from Miguel before, he was slightly concerned. "Is there a problem?"

Looking every few seconds in the rear-view mirror, and apparently trying to focus on the road, Miguel didn't hear Ron's question.

"Miguel? Is there a problem?" he repeated more urgently this time.

Miguel responded, "I'm not sure, but I think we are being followed."

"Followed?" Caroline interjected, "By whom?"

"Do you remember seeing that SUV at the stop light a few minutes ago? Oddly enough they seem to be going the same way we are… and keeping up pretty well on these small streets."

Ron and Caroline both turned around to see the SUV sweeping around the corner less than a block behind them.

"I think I can lose them though, if anyone knows these roads, I do. We aren't far from home."

But at the next intersection, they were nearly broadsided by another large black vehicle coming from the left. Maria screamed, and Miguel screeched across a small front lawn to get back onto the road.

"What's going on? Do they want our money? I'll give them whatever I have." Ron thought they must have spotted them as foreigners and wanted to rob them.

Miguel laughed. "I don't think money would interest these guys." And then he nearly turned the car onto two wheels as he navigated the next corner.

Ron felt some relief that Miguel still had his sense of humor. But one look at his panic-stricken wife wiped that away. "Just try to relax, honey, it's going to be fine, just another adventure." And to his surprise, she smiled at him.

"Daddy? What's going on?" Sophie was gripping tightly to her father's hand.

"It's alright, Sophie. They drive differently here," Ron said with a wink at his daughter.

Miguel then began talking in his native language, although it wasn't clear to Ron whether he was talking to himself or to Maria since she was not responding. Ron picked out a few words which sounded familiar like "polícia" and "casa". Then finally Maria spoke up and said "el casa". To which Miguel responded by another sharp turn accompanied by squealing tires. A moment later, they found themselves pulling into the driveway of a small house. Miguel quickly helped them out of the car, swept them into the house and shut the door.

"What was that all about?" Ron asked nearly out of breath. "Is this normal around here?"

"Normal? No. And apparently not over yet, either." He motioned for Ron to take a look out the window and pointed to the two SUVs idling in front of the house.

Rosalinda rushed to her son and took him in her arms. He quickly embraced her, and then Rosa noticed the others. "My son, you have brought guests?"

And then Miguel snapped back to reality. "Mother, you need to call the police right away."

"Why?"

But there was no chance to explain.

"Miguel, they are coming!" Ron was standing at the front door watching the action on the street.

Eight men jumped out of the two SUVs, which had parked in front of the house. They were dressed in military fatigues and carried rifles. Several went around the back of the house while two headed towards the front door.

"What is going on?" Rosa demanded.

But Miguel was running to the rear of the house to secure the back door.

"I'll explain mother, but let's go down to the cellar where we are safer," Maria said, as she grabbed her mother's hand and quickly ushered her as well as Caroline and Sophie towards the stairs, which led down into a small cellar.

Caroline called out to Ron. He replied down the stairwell that he and Miguel would talk with the men and that she should not worry.

The women were frantic. Both Rosa and Maria began to pray while Sophie cried. Caroline was in a near state of shock. She sat on a stool with Sophie on her lap and began to rock back and forth while softly singing lullabies.

Upstairs, two of the men approached the front door and began furiously banging on it.

Miguel yelled through the door at him, questioning what

they wanted. "Equação," could clearly be heard as the men continued their battering.

But then all of a sudden, shouts were heard and, as quickly as the men had arrived, they suddenly began racing back to their cars. One of the men was being dragged from the back of the house by his comrade, a streak of blood smeared across the front walk. From the house it was not at all clear what was going on.

The men at the front door also hurriedly began backing away. And from the window Miguel could see yet another group of armed men dressed in black who were the apparent cause of the retreat and who were now apparently taking control of the grounds.

Moments later, both of the black SUVs departed quickly with the second set of militia holding ground in the front yard.

Watching from inside the house, Miguel and Ron were stunned at the scene. Miguel collapsed onto the arm of the couch while Ron slumped against the door jam and wiped the sweat from his brow.

But the relief they were enjoying lasted no more than a minute. "I'm not sure we can relax just yet," Ron said, pointing to their next visitors on the way up the front steps. "Who the hell are these guys?" Ron asked with trepidation, looking at the heavily-armored figures.

Miguel jumped to Ron's side and looked out. "I have no idea, but we are about to find out."

One of the men knocked on the window pane of the door and motioned for Miguel to open up. They were not aggressive like the others, but Miguel stood firm and asked them who they were and what they wanted. In response, they produced identification, which showed that they were Papal Security, the Gendarmerie Corps of Vatican City State.

Miguel laughed. "Papal Security? That's ridiculous! Get out of here. The police are on the way!" An empty threat, since Miguel knew that no one had had a chance to get to the phone.

But Maria had come upstairs and heard the exchange. "Miguel, it is true what they say."

He looked at his sister. "Maria, what are you talking about? They claim to be the security of the Pope!"

"Yes, they are. And they are on our side."

She then briefly explained about their visit earlier from the representatives of the church.

"I find it hard to believe that someone from the Church would chase us through the streets like that," Miguel answered.

The gendarme at the door had heard the exchange. "It was not us who chased you. We are here to protect you."

Maria was sure she could trust them and that seemed to be enough for Miguel who did not stop her from opening the front door.

"Is everyone alright here?" asked one of the men.

Ron and Miguel assured them that they were, although they were still obviously shaken by all the action.

"There is no need to worry," said the gendarme, "we'll explain as much as we can." The man had a fair complexion with sandy blond hair and freckles. He was tall and muscular, with an extremely calm demeanor. His counterpart was similarly built, but with dark skin and hair.

Looking around the room, the fair-haired gendarme added, "Where are the other women? There is no need to hide. It's safe now."

At that, Maria called the others up from the cellar.

With apprehension, Caroline and Rosalinda came up the stairs with Sophie sobbing in her mother's arms. Maria approached her in an apparent attempt to calm her, but speaking to her in Portuguese only caused Sophie to bury her face deeper into her mother's chest.

Miguel then explained to his mother that they were his friends from the United States and spoke only English.

"I assume you know who I am?" Miguel questioned the men.

"Yes, you are the son of Senhora Rodriguez. And your American guests are the Stanley family from Rhode Island."

"You've done your homework," Ron remarked.

"Ron," Miguel said, "I guess you see now the power of my father's equation."

"The press won't leave us alone," Maria interjected, "but until now I wasn't afraid. We can't live like this."

Miguel then motioned for everyone to take a seat in the living room, where the gendarme explained that after their initial discussions with Maria and Rosa, they had decided to remain in the area and watch the house, just to be sure that nothing happened to them or the highly sensitive information.

"Thank God for that," Maria said.

The gendarme tried to calm them further and assured them that they were not the only two guards. There were others still maintaining posts around the perimeter of the property and neighborhood, and more had just been called in.

Miguel was still confused. "But then who were those other men? The ones who chased us across town and just tried to break in?"

The gendarme explained that all of the information they had gathered so far pointed in the direction of FARC-EP.

Ron spoke up now from Caroline's side. "The Revolutionary Armed Forces of Columbia? How could they possibly be involved in this?"

"Have you heard of them?" Miguel questioned.

"Yes, of course," Ron replied, as Caroline also nodded in agreement. "They often made headlines in the States due to their high-profile kidnappings. But haven't they been disassembled? I thought they've become an official political party." Ron continued, looking deeply intrigued.

The gendarme raised his eyebrow at Ron's apparent knowledge and went on to explain. "There were many dissidents. And rearmament has been taking place over the past few years."

"I wonder if they are trying to get the equation in order to sell it. To raise money for their cause," Ron suggested.

The gendarme continued to explain their theory. Apparently they believed that FARC was trying to acquire the equation for their own purposes, not for resale. They thought that if they could use it on their soldiers, to determine all of their deaths, then they would use only those men who they knew would survive a battle to be sent into dangerous situations. Only those who would survive the day would be used. Essentially, they would have an indestructible army, which would require much less manpower and armament than a traditional militia.

The room was silent. Everyone was stunned.

"But do you plan on guarding us forever?" asked Miguel.

The gendarme then motioned to Rosa, who began telling her side of the story.

"Miguel, I have promised to turn your father's work over to these people, placing it in the hands of the church where I feel it will be used best. But only after I have discussed it with you and gotten your consent."

"You don't need my approval, mother. Papa left his work under your care. It's ultimately your decision."

"But it would make me feel better if I had your support and blessing." She hesitated a moment and Miguel could see tears forming in her eyes. Then she said in a broken voice, "You are Eduardo's son."

"Alright," Miguel answered as he nodded approval, "if that's what you want, we'll talk it through one more time. Just to be sure."

"These people will be back," the dark-haired gendarme warned. "In the short period that the equation has become public knowledge, its existence has spread like wildfire, and the potential uses by people who would like to get their hands on it is just as extensive. People are creative, which can be dangerous if not kept in check."

"And who knows who else is out there looking to get this information." The other gendarme added. "We need to move on this fast. Your lives may depend on it."

Chapter Twenty-Seven

After another exhausting day, Amar Kumar pulled into his driveway, looking forward to putting his feet up, watching a soccer match and enjoying the take-out that he'd just picked up.

As he fumbled for the house keys in his pocket while trying to balance his dinner, he noticed the front door was slightly ajar.

He froze.

Was it was possible that he didn't shut it tightly that morning? That must have been the case, especially with all that he was going through lately.

Pacified with his explanation, he pushed open the door and stepped into chaos.

"Another call from our Professor Kumar? Two in two days. What does he want now?"

"He didn't say, Nick," his secretary replied. "He just said that it was urgent. Should I tell him you're busy?"

"No, no, put him through," Nick said, then waited for the ring. And when it came he snapped it up with a curt, "da Silva."

"Inspector, this is Amar Kumar, from the university."

"Yes, Professor, how can I help you?"

"I'm calling from outside my home. I've just arrived and it is completely ransacked."

"Okay, Professor, I will send out a squad car to have them check it out—"

"But, you will not come yourself?"

Nick sighed. "Professor, I have a schedule full of *homicide* investigations. Breaking and entering cases are not my responsibility. But I will send someone quite capable, I can assure you."

Amar began to panic. "No, no! Inspector, you must come yourself. This is related to our discussion yesterday and the deaths of my professors. I may be next!"

"Well, if that turns out to be the case, then I will get involved, but until—"

"But it is the case!"

Placing a little distance between the phone and his head to save his eardrums from the caller's shouts, Nick calmly replied, "And how can you be so sure?!?"

"I am sure. There is evidence here. You must see it to believe it."

Nick sighed, then promised he would be right over.

And just when he thought the case was wrapping itself up! There was no evidence of foul play in either of the professor's deaths, so they were both ruled 'death by natural causes'. Now, of course, Professor Montagne's death may have been avoided had he not been strapped into a straightjacket at the time, but who knows? If that crazy equation has any validity to it, maybe he would have been hit by a bus, or attacked by an angry student, and died just the same. But now this. It seems as though someone wants to get their hands on that equation.

And not surprisingly. They could reap millions by telling the rich curiosity hunters how much time they have left on the planet.

But would he himself want to know? Even if he did find out he'd live to be 90, what if he were in good health and enjoying life at 90 with the grandkids (or great-grandkids), how would he feel then? Just waiting for that day, that moment? But on the other hand, if he did find out he was going to live that long, he could live life worry-free until then! Or perhaps, he'd get the results which said the opposite, that he had only another year. Would he be able to enjoy it, knowing that the clock was ticking? Wrap up loose ends and go in peace? He doubted it, rather, he'd probably live in self-pity, a burden to his wife and boys.

Well, guess that decision is made. No equation for Nick da Silva. Not that he really had the choice. If this equation is being sought after by people willing to murder, then it certainly won't be landing in the hands of the public masses anytime soon. He was certainly looking forward to see how it all turns out though.

He put his foot on the accelerator.

Forty-five minutes later, Nick and Tomas pulled onto Kumar's street followed by two squad cars. The professor was outside and clearly anxious.

Two officers were sent to search the property and two stayed with Nick and Tomas.

"Now, what's this all about, Professor?" Nick asked.

Waving them on to follow, Professor Kumar started walking toward the house, "Come. I'll show you."

Nick and the others followed him through the small, but

carefully tended garden and onto the front steps. Nick sent the two officers in first to be sure an intruder wasn't still there. After only a couple minutes one of the officers opened the door. "All clear, Inspector."

Kumar led Nick and Tomas into the front room. There were things strewn everywhere, and the floor was covered in what appeared to be a mountain of paperwork.

"Ok. Start writing up a B&E," Nick directed his officers.

"What's a B&E?" Kumar asked.

"A breaking and entering."

"But...but...it's more than that. These people have to be found!"

"We'll do the best we can Professor, but B&Es are not top priority, especially if no one is hurt." Nick picked up a few papers that were under his feet and placed them on the nearest chair. "And I'm really not happy about having to come out here for this. I thought you said that this was related to—"

"But it is Inspector! I haven't show you yet. Come with me." And with that he gestured toward the adjacent room where a wall was vandalized with red spray paint. The neat script read:

NO EQUATION = YOUR DEATH

Nick couldn't help himself and let out a laugh.

"Do you find this funny, Inspector?"

"Well, actually, I do."

"My life is at stake here!"

"Okay, Okay. So then tell me, how are you involved in this? The equation belongs to your deceased colleague's widow, not you."

"Not necessarily."

"Enlighten me then, Professor."

Kumar then explained how the University Regents believed that they own the property rights to the work of all of their employees.

They walked outside and Kumar lit up a cigarette. Nick refused the offer of one.

"I still don't see your involvement to the point that someone would want you dead," Nick said.

After a deep sigh, which seemed to resign him to tell the truth, Kumar then explained about the barrage of telephone calls he'd been receiving from interested buyers looking to take the equation off his hands. This was the apparent result of him telling certain members of the press that any and all research work of his staff was property of the Math Department, thus implying it was under his control. He then went on to explain about the phone call from WP Pharmaceuticals and the forced offer of an eight-figure sum.

Nick quickly started adding up zeros. "Hmmm....ten million or more. That's quite impressive. I'm sure you could really use that money at the university, especially now that you've got a man-power shortage."

The cold look from Kumar told him that his last comment went too far, so he tried to reel him back in. "Alright then, if this pharmaceutical company is going to buy the rights to this work, where's the problem?"

"It's complicated."

"I have time." Nick said and took a seat on the front step while Kumar lit another cigarette. "Those things will kill ya, you know."

"Yes, maybe." Kumar mumbled as he inhaled deeply on

the cigarette then crushed it under his foot and continued. "I went to see Senhora Rodriguez. She wasn't surprised by my visit, seemed to have been expecting me."

"Really, why is that?"

"It's just what you said before. She has the equation, at least the actual software and hard copies."

"And let me guess... she told you that they belong to her and she wouldn't give them to you, right?"

"Right. She practically threw me out of the house."

Nick had to force himself to suppress his laughter this time, thinking of the stout gray-haired woman throwing the professor out of her house. "Seems to me that this may be a case for the courts to determine who has the ownership rights."

"I don't have time for that though! A representative from the pharmaceutical company is coming tomorrow morning to close the deal."

"You can't sell what you don't have." Nick quipped with a smirk.

Chapter Twenty-Eight

Nick's curiosity got the better of him and he decided to drive by the Rodriguez home. Professionally speaking, he had every reason to stop by there, considering the break-in at Kumar's. But personally, he wanted to know what Senhora Rodriguez was planning on doing with her equation, probably worth millions of dollars, and whether or not she even knew the value of it. He was followed by another squad car for back-up and when they arrived he was slightly surprised to see that everything looked peaceful. No cameras, no apparent press. Although there were several cars in the driveway and a couple more on the street. Motioning for the other officers to remain where they were, he went to the front door and rang.

Maria answered. That explained one of the cars in the driveway.

He was welcomed in and found the house full of guests, all packed around the table, which was loaded with food.

Maria led him to where his mother was sitting. Rosa began to stand, but Nick begged her to remain seated. At this point he felt a little embarrassed that he had intruded on their meal, especially since he didn't have a real cause to be there.

Just then Maria placed another chair at the table and asked Nick to join them. He kindly accepted, since it would have been awkward otherwise. He quickly radioed Tomas waiting outside, and told him he was free to go back to the precinct with the other officers, while he himself would be staying a while.

He was seated next to a familiar face. It was Ramon.

Introductions were then made to the rest. Nick had previously met Miguel, just after his father's death. Then there was another couple that he had not met, from the United States, visiting with their daughter. Apparently, they were friends or a colleague of Miguel, although the explanation of their presence was a bit vague and didn't explain all the high-end vehicles outside.

The atmosphere was friendly but not overly animated, as most large family gatherings usually were. Nick attributed that to not only a language barrier but the fact that their patriarch was missing and the wounds were still healing.

After they were finished and Maria and Caroline were clearing the dishes, Miguel finally asked Nick the purpose of his visit. He then relayed what had happened at Professor Kumar' house earlier that day.

"Is Professor Kumar hurt?" Rosa asked with alarm as she heard only the tail-end of Nick's explanation. "Please don't tell me that he has died too!"

"No, no, Senhora, he is fine, just a break-in at his home. He'll be staying with some relatives for the time being."

Miguel spoke up. "I have a feeling that you are going to tell us that the break-in is somehow related to my father's work, aren't you?"

"I am not positive," he hesitated, "but yes, I believe that to be the case." Nick accepted a cup of coffee from Maria and added milk. As he began to stir the coffee, he continued. "I wasn't sure what I was going to find here tonight. I was afraid that the same thing may have happened here, too. I am relieved to see that everything is alright, but I would like to put some security around the house for the time being, just to be sure."

Nick then noticed sly smiles on the faces of the rest at the table. "Have I missed something here?"

Miguel explained. "We already have security. And we have already had an attempted attack."

"What?!?" exclaimed Nick. "Did you call the police? What happened?"

Miguel explained about the original visit from the Cardinals, the car chase, the attack on the house, and the discussions with the Vatican's Gendarmerie Corps.

Nick was speechless.

"It's a lot to take in," Ron added. "I feel like I'm in an action film."

"I'm sorry friend. I never would have brought you and your family here had I known you'd be in any danger," Miguel said.

"Well, now that the danger is seemingly over. I can say that it was really exciting. Although... no one is going to believe me at home!"

They all laughed. But Nick had to bring the topic around to the equation again and thus back to the serious nature that it was. "May I ask, Senhora, what will you do with your husband's work? I don't think you will be safe if you keep it here with you. And I think you are aware of its monetary value?" Nick couldn't help but think how much Senhora Rodriguez could do with potentially millions of dollars. But perhaps, he thought, looking around her happy home, she is actually better off without it.

Rosalinda was thinking the same thing, about her home. She saw the way Maria and Ramon were exchanging whispers and brief touches, she thought about Miguel's wife Katie and her unborn child as well as their small children, and she couldn't imagine life any different, except of course, if Eduardo could be there with them.

"I thank you for your concern, Inspector. My children and I have made our decision. We will eventually have it destroyed."

"Destroyed? But why? You could likely earn millions from the sale of it."

Looking up as if to the Heavens, Rosalinda smiled. "Inspector," she began, "I have seen anger, greed and violence, which have stemmed from the desire to control my husband's work. This is not what he would have wanted." Just then Caroline walked back into the room and sat down at the table. Rosa gave her a broad smile and a tear escaped from her eye before she continued. "I know my husband would have supported our decision, but before his work is destroyed, we plan to use it for the purpose of good… at least I pray the outcome will be positive."

Of course Nick did not understand completely what she meant, but it was not really his business. He left the house with the feeling that they were safe. And that was his job.

He went home to his wife and boys, giving each one a hug and a kiss, to the delight of his wife and dismay of his boys. Then sitting at the dinner table, for his second meal that night, he asked, "So, what kind of dog are we getting?"

Chapter Twenty-Nine

The aircraft was magnificent.

A customized Boeing 747-200, the same type of aircraft flown by the President of the United States.

Sophie was sleeping on a recliner by a window, a blanket draped over her, but Ron and Caroline were too wired to sleep, excited to be on the next step of their adventure.

"I can't believe we are flying on this alone!" Caroline said.

Ron smiled. "Well, we are not exactly alone. There must be at least twenty airline staff on board."

Caroline leaned back and sighed. "I could get used to this."

Me too, Ron thought, me too.

Rosa had convinced the gendarmerie to allow Ramon to use the equation on Caroline. Only after Caroline had her answer would Rosa hand the equation over to the Vatican.

Naturally, Ron and Caroline had to fly home to gather all the necessary information that Ramon would need to make the calculations. So, since the Vatican did not want to waste time, they offered to put Ron and his family on a flight that same evening so that they could gather the required data as soon as possible.

Ron had a list seven pages long of numbers that they needed to collect. But they felt as though it was the key to their future, rather than a death sentence, which in reality it was, whether now or decades in the future. But as he reflected back on the last few days, he suddenly felt exhausted. Ron took his wife's hand and, looking into her face, he swore he had never seen her appear so beautiful. "You are so brave."

She smiled. "So are you," she replied.

"Me? Why me?"

Caroline sat up straight and looked squarely at her husband. "You are the one who brought us here. You are the one who gave us this chance." Then her face turned grave. "Although, we could find out next week that I will not survive the next year."

"Don't talk like that, we don't know—"

"Shhh... Ron, it's alright. I feel at peace. At least we will know. We will know which decision to make, and not have to second-guess ourselves." Ron opened his mouth to speak, but Caroline continued. "Right now, my chances of survival are very low. We both know that. At least, if we find out that I have only a short time, then we won't waste it with treatments that will make me feel worse or worry about getting pregnant with a baby which may never live." Sophie stirred in her sleep but Caroline went on. "You have to promise me something, Ron."

"Of course, Caroline. Anything."

"If I'm not going to make it, you have to promise me that once we wrap up loose ends, we will enjoy every day that I have left. We won't burden ourselves with misery for the lost time that we will never have, but rather be thankful for what we have been given." At this, Sophie woke from her sleep and came over to her parents. She curled up in her mom's arms and closed her eyes once again. Caroline did the same. And it was clear that she was living in the moment. And enjoying it, too.

"I promise, Caroline, at least I promise to try," Ron whispered. Then he too closed his eyes and soon drifted off to sleep.

Chapter Thirty

Nick was early to work the next day, but before he could even sit down and browse through his in-bin, his door flew open.

"Ramon has been kidnapped!"

Nick almost didn't recognize Manuela as she burst into his office half out of breath. She was shaking and extremely agitated. Esposito came in right on her heels and shut the door behind him.

Nick sat up straight in his chair while Manuela paced. Looking at his boss, he said, "Shit. What happened?"

Esposito took a seat. "Apparently Manuela's nephew Ramon had been at the house of that professor who had died a few weeks ago, the one whose files we decoded."

"I know, I saw him there last night" replied Nick.

"You did?" Manuela ran over to Nick and grabbed him by the shoulders. "Was he all right?"

"Yes, he was fine." Nick pointed to the other chair in front of his desk. She sat down. "Now calm down and tell me what happened."

She took a deep breath and continued with the story. "My sister, you know, his mother, called me last night after he didn't come home, that was about 11 p.m. She said that he had been at the professor's house. She thinks he's got something going on with the daughter, Maria."

Esposito leaned back and crossed him arms. Clearly he'd heard this before.

"And did you try calling the Rodriguez home?"

"Yes, my sister did last night before she called me. They said that he left shortly after nine and that he was going home. She also called the university but he wasn't answering the phone at the lab either. But that is all irrelevant now," she said as she banged her fist on the arm of her chair.

"How is that irrelevant, every clue could lead—"

Now Esposito interrupted, pointing to Manuela. "She got a call a few hours ago from her nephew—"

Manuela broke into the conversation again. "He sounded bad, Nick. They must have beaten him, his words were all slurred."

"Okay, don't worry, we'll get him. Just settle down and get the story out. What did he say? Why was he calling you?"

"He said that he needed the computer, the one from the second dead professor."

Nick was right with her. "The one that had the functioning equation on it."

"That's right."

"Shit." Nick put his head in his hands.

Manuela's heart rate picked up again. "What?!? You look like you know something."

Almost to himself Nick mumbled, "I didn't even think of that... of course Ramon would be at risk."

Esposito leaned forward. "You want to tell us what's going on here, Nick?"

So Nick relayed everything that he had learned the night before from Miguel and his family about the Vatican's involvement and the FARC attack.

His boss looked like he was about to explode. "And didn't you, at any point, think it may be pertinent to report any of this to the authorities, namely, me?!?"

"Shit, boss. The situation was under control when I got there. The family didn't want to report the attack because they didn't want to draw more attention to themselves. But just to be sure, when I left I stationed two officers at the house to protect the family."

"Under whose approval?" Esposito boomed.

"Under mine! I don't need approval to assign a couple of officers on watch. Or do I?!?" Nick was getting angry under the attack from his boss.

Esposito let out a long breath. "Okay, you're right, Nick. You don't need approval for that. But damn it, the Pope's forces here and a FARC attack, too?!? This is way over your head and mine for that matter. And you were completely out of line to keep all this information to yourself!"

"It would have all been in my report anyway," Nick added in his defense.

As though Manuela had suddenly reappeared in the room she brought the focus back to point, "Hey, you two can work out any problems that you have later on, right now we need to get my nephew back!"

"She's right. We'll talk about this later," Esposito said.

"Fine," Nick replied. "Go on, Manuela. What happened next?"

"Well, when I got off the phone I came right here to

collect that computer from the evidence room where it's been stored."

Nick was still clearly frustrated. "Manuela, you can't wantonly hand over evidence to a revolutionary group just because they want it!"

"This is my family, Nick, I would do anything for them! Especially Ramon, he's just starting out and has everything to look forward to in his life."

"Okay, Manuela," Esposito said, "we'll address this later, too. Go on."

"I got the computer and left. I met them where they told me to, in Villa-Lobos."

"The park?"

"Yes, they met me at the park, on one of those side streets. I demanded to see Ramon, but he wasn't there with them. So I told them they weren't getting the computer until I got Ramon." She then put her hands on her face and started to wail and snort quite loudly.

Esposito and Nick exchanged looks.

"They held her at gun point and took the box," Esposito continued. "They told her they weren't done with her nephew yet. Then they took off."

"Oh man, Manuela, you all right?" Nick asked.

She nodded with her face in her hands but the high-pitched crying continued. Turning back to his boss Nick asked, "What else do we have? Auto make? License? Physical descriptions?"

"Yes, we've got pieces of the puzzle but no definite leads. We've already got someone working on tracking down the vehicle."

Nick got up and grabbed his coat. "Okay, give me what you have and I'll follow up on it."

Urso stood up too and approached Nick, up close he towered over him. "What are you planning Nick? I don't want any renegade moves. Let's do this by the book."

Nick hesitated. "I can't do that Urso, not just yet. Get that info and get in the car with me if you want to watch my back."

Esposito didn't hesitate. Somehow he wanted to be there for this one. He quickly went to his office where he got the report, his weapon and at the last second grabbed his armored vest, then he hurriedly followed Nick out to his car.

As they drove, Nick wondered how to bring up the topic about their disagreement earlier. He'd noticed that his boss had been really edgy lately, whereas he used to be a very free-handed supervisor. Not to mention the fact that they'd worked together for years and had come to know each other well enough to trust one another. What was going on with him? Mid-life crisis already? Or maybe problems at home? Just last week he mentioned something about being frustrated about guys joining the force for the wrong reasons, as a power-wielding position or something. Then, just as he was about to ask what was going on, Esposito spoke up.

"Look, Nick, I'm sorry about the outburst before. There's, well... there's something going on with my brother..."

"Forget about it. But your brother... I hope he's OK. Is he sick or something?" Esposito's brother was also in the precinct, he was several years his junior, and was expected to work his way up quickly like his big brother.

"No, I wish."

"What? What's that supposed to mean?"

"No, that's not what I mean. I don't wish him harm. It's just that I wish it wasn't what it is."

Nick raised his eyebrows. "You've lost me now, boss."

Esposito sighed, then let the story out. "We've had an investigation going on for some time now, a construction fraud scheme."

Of course Nick hadn't heard anything about it since it was not in his department, but then again, it wasn't in Esposito's either. "How are you involved with that?"

"I'm not. They've simply notified me. My brother is heavily involved. He's going to be indicted."

"Oh, that's rough. I'm sorry to hear that."

"Yea, thanks." Shaking his head, Esposito continued, "This is going to kill my father." His father had been a police officer for forty years until his retirement a few years ago. He had been highly respected by everyone on the force as well as the local politicians, businessmen, and justice department. His reputation was completely unblemished.

Nick didn't know what to say, he was not the type to console, but he thought about Esposito Sr. and how he would feel. "Your father is a strong man. He'll get through this embarrassment... just like you eventually will."

"Yeah, I guess..."

"And this gives you all the more reason to hold up your reputation, by keeping up the good work," which was, in effect, a hint at not screwing things up by losing his patience with his subordinates. Then, trying to lighten it up, Nick countered, "It could be worse... your brother could be sick."

"Oh shut up, you idiot," replied Urso suppressing a smile.

Mr. Jonathan Boyd was a tall, thin man with graying hair

and a long, bony nose. With his scholarly spectacles, he looked more like a college professor than a pharmaceutical company representative. He fit right in at the university.

When Professor Kumar returned from his lunch the man was waiting alone for him in his office. "May I help you?" Kumar asked, wondering if it was one of the new candidates for his open teaching position.

Mr. Boyd stood and shook hands with the professor. Kumar was a little startled at what a strong grip the man had considering his rather awkward appearance.

"Of course you can help me, Professor. We spoke yesterday. I'm John Boyd. You have something for me, I presume." His face showed no emotion as he spoke and he stood straight and still, waiting for a response.

Kumar was speechless. Inspector DaSilva had promised him yesterday that he would handle the situation. That he would visit with Eduardo's widow and perhaps even subpoena the equation if necessary. The Inspector had also said that he would contact the university Regents himself and have Kumar removed from any involvement.

At first he didn't agree with the Inspector, since he enjoyed being in the limelight, but then he consented and was actually relieved at not having to worry about having his home being broken into again.

"I... I... Have you spoken with the Regents today?" Kumar asked.

"No, of course not, that part of the arrangement has already been completed, and I don't make a habit of having to repeat myself." Placing his briefcase on the desk and popping it open, he continued. "I'd be pleased to make this as brief as possible. All hard copies of the equation and notes as well as electronic copies shall be placed in here." The same earnest face did not falter. "Immediately, please."

"But I don't have it. And it may be a while until I do. There may even be a court case involved."

"I'm a busy man, Senhor Kumar. I will wait here one hour, and then I expect to be leaving with what I came for." At that, Mr. Boyd shut his briefcase, walked over to a hard-backed chair by the door, and sat down. With his briefcase on his lap he cocked his chin upward and settled down to wait.

They were quickly out of the city and by now Nick knew the way without navigation. "So, where are we going?" Esposito asked.

"I'll give you one guess," Nick replied.

Actually Esposito knew himself before he even asked. "The site of the FARC attack last night?"

"You make it sound so insidious. Nothing happened."

"Yes, we were lucky that time," Esposito said. "But FARC is ruthless and doesn't make many mistakes. Hopefully we'll get Manuela's nephew back... alive."

"We will," Nick reassured him. "And that's why we are headed over to the Rodriguez house. That's where these gendarmes are staked out. At least that's the only place I know about. Although, by now they've probably got a secret high-tech headquarters somewhere downtown." Nick sighed when he had to stop for a red light. "But they most likely have more info than we do and from what I've heard they'll do just about anything to get that equation in their hands, so I'm betting they'll help us out in getting Ramon and that computer back."

"I hope you're right on this one. It hits close to home," Esposito said, clearly referring to the family of one of his officers.

"How much firepower do you have on you?" Nick asked.

Esposito shook his head and sighed. "Just my standard issue. I'd planned on spending the day doing office work."

"Maybe that's the problem boss."

"What problem?"

"Yes, well, I get the feeling that your frustration is not just 'cause of your brother. How about getting back on the streets?"

"And take away your fun? Nah, I'm happy where I am, although I have a feeling today could be interesting."

"Hmmm, interesting? I wouldn't put it quite that way, but for some reason I also have a feeling it's going to be a day for the books."

They approached the Rodriguez residence and parked on the street behind a row of vehicles. Instead of going to the door, they went directly to one of the other cars, and were pointed out where they could find the commander of the gendarmes. Recent events would most certainly be of interest to them, considering that FARC had out-tricked them and gotten their hands on a functioning equation, in addition to probably the only person alive who could operate it. Nick was curious to see how much they already knew.

Nick wasn't disappointed. Of course they knew about Ramon's abduction, but weren't so concerned since it was just Ramon and not the equation itself. Nick relayed the news that FARC had contacted the hostage's aunt and met with her. This was new information to them, and it got their attention. The gendarmes went immediately into action. They eagerly devoured the report Urso showed them and after a few calls, two of the gendarmes were in their car ready to go. Nick was still standing on the sidewalk next to their vehicle. "Where are we going?" he asked.

The two gendarmes looked at one another, then the passenger replied. "Thanks for the tip guys, but we've got it covered." And they started the engine and pulled away.

In about an instant, Nick and Esposito were in their car too and right behind the gendarmes, who must have known that they were being followed, but they didn't seem to mind. They were driving fast, but obeyed all traffic signals. Nick was frustrated with the pace, but didn't dare put his flashers or siren on so as not to scare away the suspects on arrival. He wasn't sure where they were going or how far away, so he just stuck as close as he could so he wouldn't lose sight.

They headed north-east for about fifty minutes and ended up on the outskirts of a deserted military settlement, an ammunitions depot, near the coast. It was an immense piece of land, but the buildings and roads had deteriorated over the years from disuse.

Nick could just see a series of Quonset huts in the far distance when the car carrying the gendarmes in front of them turned into a dirt lot, which was partially protected by some scrubby creosote bushes. There they met up with four other cars.

The two gendarmes got out and immediately approached Nick's side of the car. He jumped out before they could get there.

"There is nothing you can do here." The gendarme said. "You must leave."

"Like hell." Esposito boomed as he came around from his side of the vehicle. "This is our jurisdiction, so you'd better get used to the idea of our involvement."

The presence of a marked police car drew the attention of the other gendarme agents, and they came over as well. They were all informed that these two Sao Paulo officers were the ones who had supplied them with the new information that an operational equation was now in the hands of FARC.

The gendarmes were discussing the situation when Nick butted in, "C'mon guys, we have no time to waste."

Then the man who appeared to be the superior nodded and turned to his men saying simply, "Brief them quickly and then let's go."

Chapter Thirty-One

The luxurious aircraft that had carried them home departed immediately after refueling. Ron and Caroline wouldn't be travelling back to Brazil, at least not now. They simply needed to gather the required information for the equation and email it to Ramon. A few hours after that, they should know their fate.

After a very short night's sleep the entire family met at Ron and Caroline's at 2 p.m. Caroline's family members were told what to bring: medical records, birth records, passports new and old documenting travel, photo albums and yearbooks as memory joggers, and most importantly, a clear mind. Ron had his laptop on hand to record the data and for quick queries.

Even Ron's parents came so that they could be a part of it, although they weren't expected to contribute much to the 'statistical gathering', as they called it. They were mostly busy taking turns looking after Sophie while Ron organized the information.

They all knew vaguely why they were there, but it wasn't until they were all seated and Sophie was safely out of earshot that Caroline began explaining everything in detail that they'd been through over the past few weeks: the doctor's visits, therapy options, their trip to Brazil and, finally, the equation and why they planned to use it.

Of course, they left out the part about the FARC attack, although later over dinner Sophie started recounting stories about it. Ron and Caroline laughed it off as though she had a terrific imagination. Both had agreed that some things were better left untold, for the time being anyway. They planned to take Sophie to a counselor once everything had settled down. She seemed fine after the ordeal, but they didn't want to take any chances that she was repressing the traumatic memories.

Ron and Caroline talked and talked. And with so many questions from the family, just the background part of the conversation took almost two hours, at which point they hadn't even begun on the data collection.

But then it was finally time.

"So, what are the latitude and longitude of Providence?" Caroline began.

Without missing a beat or even looking at his notes, Ron answered, "Latitude: 41 50'; Longitude: 71 24'."

"How the hell do you know that?" Caroline's sister, Jess, asked.

"I've had this paperwork for a day already, and some of the facts were easy for me to gather."

"Okay, next," Caroline continued. "Height above sea level?"

"What? The city of Providence or the hospital you were born in?" Jess asked again.

"The hospital."

"Which hospital was it?" Ron asked.

"You don't even *know*?" Jess asked.

Caroline came to her husband's rescue. "Same place you were born, 'Lying In'. Although it's now called 'Women and Infants'."

"Oh. My. Gosh. That's so funny. Lying In?!?" Jess mocked. "I'll bet a man came up with that. How original!"

Caroline's father spoke up. "Okay, let's get serious. We have a lot of work to do and it's already almost time for dinner."

Caroline's mother then added. "You're right, so where were we? Oh, the hospital or the city?"

Caroline answered. "Well, Ramon said that everything should be as accurate as possible. So, I guess the hospital. Can we find out its height above sea level?" Then after a moment's thought, "Actually, it's only a stone's throw from Narragansett Bay, so it must be just about at sea level, and I think the longitude and latitude for the city itself are close enough."

"Okay, then let's say 50 feet," Caroline's father said. "But what if you were born on the tenth floor instead of the second? There's a difference of... I don't know... another 80 feet maybe?"

"A figure of either 50 or 100 feet may be good enough, especially when you are comparing with someone born in the mountains at 10,000," Ron interjected. "But, we have to think about time here, too. We could spend weeks on this or we could get it done tonight."

"That's true," Caroline agreed.

"We just can't take forever to get this information together," Ron continued. "I mean, we need to be as accurate as possible, but if we can get an answer out of this magic equation with an accuracy of say, within six months, then we have what we are looking for."

The dull reality of the game sunk in and there was silence for a few seconds, until Sophie ran in and announced that dinner was served. This actually meant that the pizza had arrived and Ron was to pay the man at the door. In the intense conversation, he hadn't even heard the doorbell ring.

They took a break for dinner. The mood was hopeful but serious. Sophie was brought to bed early since she had almost fallen asleep on her pizza. She was feeling the effects of the excitement and travel. Then they started up again with their work.

"Height?" asked Ron.

"Five feet, five and a half inches," Caroline responded.

"Actually we need your height in centimeters." Ron took out a piece of paper and scribbled down some numbers. Then he tapped a few keys on his calculator and announced, "166.37 centimeters."

"Do all the values have to be in metric?" Ron's father asked. "If so, we need to change the figure for altitude."

"Oh, yeah, I think you're right," Ron said. Then he recalculated the value and made sure there weren't others that he'd missed.

The list went on: blood type, number of countries she had travelled to, if she was afraid of heights or claustrophobic, and whether or not she could swim!

They began to get frustrated as they mucked through the subject of chemical exposure, covering use of hairspray and other aerosols, proximity to high tension wires (less than 100 meters; 100 to 250 meters; or more than 250 meters; and for how many years), diet, exercise, and whether or not she meditated. The final series of questions were about childhood diseases and other illnesses. How many times had she had the flu before she was ten years old? That was impossible to say for sure, so they had to guess.

There were hundreds of such inputs and the whole family was exhausted when they finished the list. Ron looked at his watch, midnight, exactly. "Sao Paulo is two hours ahead of us, so I'm sure they are fast asleep. I'll summarize all this and send it out first thing in the morning."

The family slowly got up and Caroline hugged each one of them as they said their tired goodbyes. Ron and Caroline promised to call everyone as soon as they heard something back from Brazil.

Good or bad.

Chapter Thirty-Two

Nick and Esposito were given a very brief run-down of the situation and had their assignments, which was essentially to stay out of the way, but at least they were on site. It was a clear, dry day and the sun was high in the sky. Visibility was impeccable, which could help or hinder them, depending on who was watching who.

There were eight gendarme members who were closely huddled around the hood of one of the cars, carefully reviewing a map of the complex.

Nick and Esposito had their eyes trained on the Quonset huts.

They were waiting for the signal when Esposito whispered, "Are we getting in over our heads here, Nick? We have no body-gear other than vests and we're only minimally armed."

"We're just back-up," Nick said. "We'll be fine."

"Maybe," Esposito said, "but I need to be sure about that." He then returned quickly to his vehicle and radioed the station.

Two minutes later the head of the gendarmes announced, "I have them in sight."

Immediately Nick peered towards the Quonset huts to see if he could spot the kidnappers, but he saw no one. Then he followed the line of sight of the leader and noticed that he was not looking towards the huts, but rather, in the air, where there were four parachutists headed in their general direction.

Nick and Esposito had been told that a special assault team would go in first. This must be them, Nick thought.

Since the landscape was flat and bare, any approach on foot or by car would be spotted at some distance, thus the initial vertical approach. Nick was impressed.

Apparently, the gendarme team had been following the movement of the FARC members for the past several days and had known about this hideout. Although, how FARC had managed to kidnap Ramon and, more importantly, acquire the computer with a functioning equation without their knowing about it clearly had them bothered. Nick wondered why the gendarmes hadn't known that the police also had possession of a computer with the equation. They seemed to know everything else. He started to feel relief in the fact that his boss had just called in for support.

Now the action was about to start.

The parachutists would land, enter the building where the militia were suspected to be, and make the initial assault. They would try to apprehend the militia or, worst case scenario, act as a diversion until the additional back-up coming in by road could get there.

Nick and the others were instructed to wait thirty seconds after the landing and then move.

They all packed into three of the unmarked vehicles, and the approach took about sixty seconds. When they arrived at the Quonset huts, the gendarmes burst out of the cars and took off running.

But they were only met with silence. Nick was confused. He had expected an onslaught upon arrival. But the leader motioned for the rest of them to maintain their positions. They were to remain by the cars, so they stayed put behind the vehicles. A half minute later, still silence from within.

Why was there no yelling? No gunshots?

Nick's instincts told him that something was wrong.

The sudden, powerful explosion proved him right. It was a blast which blew the roof off the building and sent him and Esposito sailing about ten feet backwards. Both were coughing and temporarily blinded by flying debris and smoke.

Then Nick heard his boss call out. "Nick, Nick, you all right?"

"Yeah, I'm Okay. Where are you?"

"I can see you. I'm about five meters, two o'clock."

Nick lay for a moment on the ground, trying to get his bearings. The building was now on fire and his first thought was Manuela. If her nephew was in there, then he and Esposito would have to be the ones to give her the bad news.

No one could have survived that blast.

The ground was rough, dry gravel. Nick rolled to his left and crawled the few yards to where his boss lay. Urso was clutching his left arm. "You hurt?" Nick asked.

"I think I caught some debris, but I'm ok." He looked around at the scene. "We've got to get out of here."

The cars they arrived in were in flames. Esposito was slithering on one side, trying to protect his injured arm by tucking it into his body as close as possible. They stayed on their bellies and began moving in the direction of the cars stashed further away, a distance of about half a mile. It looked impossibly far. Nick decided to risk it and stood up. He was about to start running when he heard a commotion from behind. He and Esposito, who was also now on his feet, both stopped in their tracks.

They turned around to see two of the gendarmes running from the backside of the building and also heading towards the stashed cars. In an instant, they heard a loud crack and saw one of the men go flying off his feet. A gunshot. The other man stood still a moment, stunned, then bent down to his comrade. But there was nothing he could do, the man had been hit in the neck and he wasn't moving. So the gendarme got up to run again, and that's when Nick saw the reflection of the scope. It was coming from the direction of a warehouse, about 250 meters away to the right.

In a heart-beat, both Nick and Esposito took cover behind the burning vehicles. There was nothing else around to protect them from the sniper.

Nick turned to Esposito, "I've got their position."

"Yeah, I see them, too. What are we going to do? If we stay here we're sitting ducks. If we try to run, we're dead."

They heard two more gunshots and assumed the other gendarme was dead too. But apparently his body gear had taken the blows, because a few seconds later he had managed to make it to where Nick and Esposito lay.

None of the others appeared to be around. Either they were still in position on the other side of the nearly decimated building, or their bodies were lying inside it.

Nick pointed out the location of their attackers to the gendarmes, but it didn't do them much good. They had no clear shots, had no long-range weaponry, couldn't see any targets anyway, and knew that they were being watched.

Their options were non-existent. They could only sit and wait for the backup that Esposito had called for and in the meantime see what fate had in store for them.

Chapter Thirty-Three

Miguel woke up and was temporarily confused. Where was he? He then realized he was in his childhood bedroom, and the events of the day before came flooding back to him.

They'd gotten word about Ramon's disappearance late the night before and he was hoping for some positive news that morning.

He got up and went into his mother's kitchen where he found her sitting at the table next to Maria. His mother's arm was wrapped around her daughter in a loving, protective gesture. Maria's face was stained by tears and her eyes were swollen and red. Clearly they hadn't received any good news. Rosa looked up as her son entered and gave him an almost pleading look, to help somehow, since she herself was helpless.

Miguel sat down and poured himself some coffee.

He took his sister's hand and the look in her eyes told him for the first time how she truly felt about Ramon. She was in love with him. He knew he had to do something to help. Although he had his wife and children back in Brasilia, his mother and sister were his family too, and now that his father was gone, he was also the head of their household.

He finished his coffee and stood up, then gave his mother and sister each a kiss, and walked out the back door.

His old Honda motorcycle was where he had left it in the back of the garage. He'd used it throughout college but left it behind when he moved to Brasilia, only occasionally starting it up when he was home for visits. After years of disuse, the motorcycle still looked amazingly clean. Even the tires were full of air. Miguel was not surprised when he kick-started it to life on the first try. His father had clearly kept it in shape. The thought of his father brought a smile to his face as he accelerated and headed towards the city.

Thirty minutes later he was talking to Nick's secretary at police headquarters. She greeted him with a flirtatious smile, though he hardly noticed. He was an attractive man and most women responded to him in that manner. It was something he was used to and attributed it to simple, friendly human nature. He asked for Inspector da Silva but was told he was not in his office. And whether she sensed the urgency in his voice or was just looking to please him, was unimportant. Instead, she offered to try to contact the Inspector on his mobile phone.

As she dialed, Miguel took a look around the office. There were several groups of desks where officers both in uniform and street clothes were busy working the phones and the computers. Though everyone was busy, the place had a sense of calm, as though everything were under control. How could that be? The Pope's gendarmerie and FARC were running rampant through the city and a young PhD student had just been kidnapped. Why wasn't there chaos and panic?

The sound of the telephone receiver clicking into its base jogged him from his reverie.

"He's not answering his phone. Can I leave a message for him?"

But before he had a chance to respond, an officer rushed up to the desk and told the secretary to call the switchboard and send as many officers as there were available to the deserted ammunitions depot as back-up for a hostage situation. She picked up her phone and glanced to where Miguel was standing, raising a hand to signal that he should wait a moment and she'd get back to him. But Miguel didn't stick around, and before she'd hung up the phone, he was already out the door.

He didn't need directions to the depot, it was the only one of its kind in Sao Paulo, and his father used to take him there when he was a kid to fly kites. The flat terrain and wind off the water made it the best place for kite flying. As he rode off in that direction now, he thought back on the one time he and his father took Maria with them to fly kites. Until then it had always been something he had done alone with his father, but this time she had begged to come and they relented. She was only about six years old at the time. The day had been windier than usual, so Miguel was excited. His kite, an aerodynamic wonder, was flying high just minutes after their arrival, while his father was helping Maria with hers. She had a very simple one, that looked almost more like a balloon than a kite, but it was pink and that was all she cared about. His father had just gotten hers aloft and handed her the grips when she suddenly let go. The kite kept sailing, higher and higher, caught up in a mighty gust of wind. The grips were kicking and dragging along on the ground behind, the string unraveling as it went. His father ran behind in his best attempt to catch it, but it soon ran out of string and the grips were airborne. The kite was gone and Maria was in tears. Miguel began to curse his sister for her stupidity, but his father gave him a stern look which told him to stop immediately. His father then took his little daughter into his arms and began to sing to her while rocking gently side to side.

Miguel watched in wonder. He had always known his mother to be the gentle nurturer, but seeing his father, a man, in that role, was something new.

Even now, he often thinks back on that day, to give him support when his own little daughter, Rosie, needs comfort. Compassion was one of the many things he learned from his father, in addition to fixing motorcycles and flying kites.

But today he wouldn't be flying any kites. His visit was for another purpose entirely.

Chapter Thirty-Four

The main entrance into the old ammunitions depot was on the south side of the complex, only about five hundred feet from the rocky coastline. When Miguel arrived, there were already four police cruisers blocking the road, another group of vehicles, which may have been unmarked police cars or perhaps civilians, were parked in the dirt lot just inside the gate.

Miguel played dumb and asked what was going on. A 'security issue', he was told, but the officers were too busy relaying information back to headquarters to talk to him for long. Inside the complex, in the distance, he could clearly see smoke and some flames. He had a very uneasy feeling.

After a few minutes one of the policemen got out of his car and addressed Miguel, "You'll have to leave, sir."

Miguel simply nodded and took off in the opposite direction from where he had come. The road was essentially a ring around the outside of the complex. The fires were off to his right and as he drove he kept trying to recall the layout of the depot. There was not much there. Lots of open space which sloped gently towards the water, no trees of any significance, and just the few warehouses and Quonset huts, but of course the entire facility was surrounded by a high chain link fence, so access was limited. Eventually, he could make out that one of the Quonset huts was on fire and there was another smaller fire near it, which for all he could tell it could even have been some burning debris from the structure.

The paved road eventually turned to gravel and Miguel knew he was almost there. Once, when he was younger, they had had a terrible bout of fall rains, and on the first sunny day after what seemed like an eternity of foul weather, he and his father headed straight out to the depot for kite flying. But when they arrived the entire main road into the complex was flooded. They couldn't get in. Seeing the disappointment in his son's eyes, his father suggested that they try to find another entrance. Of course his father probably didn't truly believe that there was another entrance, but he somehow wanted to assuage his son's disappointment. Much to his surprise and Miguel's joy, they did find a back gate, which opened onto a narrow rugged road that took them right to where they wanted to be.

And now Miguel was there again. The gate was completely gone, torn off its hinges and apparently carted away, and the road inside was in even worse shape. He considered parking his bike and walking, but then thought again about his safety and that he might need to make a quick exit. So he rambled on as best he could, weaving around the potholes, until he realized that he'd be much better off riding through the scrub grass in the field adjacent to the road.

He continued in the direction of the smoke, with all kinds of scenarios running through his mind about what he might find ahead. The police officer at the station simply said it was a hostage situation. What if it's another hostage situation, totally unrelated to Ramon? He could be risking his life for nothing. He could almost hear his wife cursing at him for putting himself in danger.

But he also had his sister to think about, so he kept going.

Then he noticed a vehicle coming toward him. It was kicking up a lot of dust and moving at a good pace despite the quality of the road. As it approached he saw that it was a Jeep CJ-7, with no cover. There were four passengers and he strained his eyes to see if one of them was Ramon. It was impossible to tell until they were right up on him, and at that point he noticed that they all had rifles and revolutionary-style combat clothes. Miguel felt a sense of déjà vu, as the men resembled those from the recent car chase. He assumed he was looking at four FARC members. One of the men saw Miguel and pointed him out. A rifle was aimed in his direction and he picked up speed, making a bee-line away from the road. He heard a couple of wild shots but didn't slow down to see if they were coming after him, he just rode with his head down for a minute or more, until he saw a warehouse about a quarter mile away. He made for the structure, pulled around the side, climbed off his bike and tried to catch his breath. A look around the corner of the building indicated there was no sign of the Jeep. A moment of relief was all he could enjoy until he realized he had no idea if he was safe, didn't know where he was, nor what he should do next.

From where he was now standing, he could see the fire much more clearly. It was completely engulfing a Quonset hut about 200 or 300 yards from where he stood. He could also see the police in the distance. It looked as though they were making their approach along the main road. What were his options? If he came out towards them, they may mistake him for one of the hostage takers, and if he waited much longer to get out of there, they may simply ambush him.

He was in over his head and he needed to get out of there. Immediately.

He turned to go retrieve his bike and then froze as he saw a back door to the building slide open. He flattened himself against the wall and slowly began inching himself backwards and around the corner.

He could hear several voices inside. They were arguing. He tried to make out what they were saying, but then the sounds were muffled when he heard an engine start. A vehicle burst out of the building at high speed, heading in the direction from which he had just come. It was a dark SUV and Miguel could have sworn it was the same as the one that had followed him through the streets of Sao Paulo just the day before. But where was the second one?

He knew now that he was in the right place, and that Ramon must be somewhere nearby.

Just then, as if in answer to his thoughts, a painful cry penetrated the silence. And then more voices. Shouting.

"We have to leave. Now!"

"And the hostage?!?"

"We're done with him. Leave him. He won't survive that beating anyway!"

Miguel heard another cry of pain and moaning, which he now recognized as Ramon. A sense of both relief and dread settled on him.

Car doors slammed and an engine started.

And, just as expected, the second SUV raced out of the building.

Miguel held his breath and waited, then slowly began inching his way towards the bay doors. He risked a quick look, but his eyes had trouble adjusting to the darkness inside the building. He saw no one. A second longer look and he found what he had come for. Ramon. He was laying on the floor, his head facing the direction of the doorway.

He must have seen Miguel.

"Socorro." Help. His voice was weak.

Miguel rushed towards him. "Ramon. It's me, Miguel. I will help you."

Ramon lifted his head and first appeared confused. Then a look of relief before he collapsed onto the ground again.

"Ramon. Ramon. Can you hear me?" Miguel shook Ramon's body as gently as he could, but there was no response. He had apparently lost consciousness. Miguel checked his vital signs. He was still breathing but had a weak pulse.

Without having any idea where the rest of the FARC militia were, or even if there were more, Miguel knew that time was of the essence and he needed to move quickly. It could easily be the difference between life and death for his friend.

Ramon's face was swollen and discolored, he was hardly recognizable, and Miguel couldn't begin to guess what other injuries he may have. But one thing he knew for sure was that he needed to get them both out of there and to medical care as fast as possible.

He lifted Ramon from under his arms and placed him over his left shoulder. Thankfully, Ramon was relatively small and not exceedingly heavy. Miguel lumbered onto his bike and, as carefully as possible, positioned Ramon in front of him on the seat. But it would be impossible to hold him upright and drive that terrain with one hand. He had to think fast.

He scanned his surroundings and saw a pile of rope nearby. He leaned Miguel over the handlebars while he went to retrieve it, tugging it to test for the rope's strength as he walked back to the bike. He remounted and then wrapped the rope around his back and then across Ramon's chest, forcing them together like two tandem jumpers. Now Miguel's hands were free.

He hit the ignition and took off, extremely unsteady at first.

Then he stopped. He wasn't sure in which direction to head. He was afraid to move towards the police. What if they mistook him as one of the hostage takers, caught red-handed with the hostage? But he was even more fearful of meeting up with the militia. So, he chose the police, at least they'd give him a warning before they started shooting.

Riding across the dry, rocky terrain with the added weight, and the extreme care it took to balance Ramon, it was almost impossible to make progress. Having no better options, he decided to head straight toward the Quonset hut that was on fire, figuring that the fire department would soon be heading there if nothing else.

In the distance he could make out several police vehicles also moving in that same direction, toward the fires. And as he got closer he could make out that the fire near the building was in a vehicle. Or maybe two vehicles?

He had a very anxious feeling. But with the police there, they would be safe.

Or so he hoped.

Chapter Thirty-Five

Esposito tucked his phone into his pocket and assured Nick that back-up was on the way.

But Nick was uneasy. He was laying on the ground between two burning vehicles and a burning building, not to mention being in the sights of a sniper. He was completely helpless and began to get sentimental. Thoughts of the late dinner with his family the night before came to mind and the promise to his boys about getting a dog. In the morning light he had almost regretted making that promise and wondered how he could get himself out of it. But being in his current situation put a whole new perspective on things, one of which was that he would give his right hand to see his family again. He knew, once he got out of there, they'd be getting that dog after all.

Almost an hour had passed. The fires began to die down. They had heard some vehicles in the distance but having no idea if the sniper was still in position, they were in a lockdown. It wasn't safe to move. He began to feel very tired but forced himself to stay awake by memorizing his surroundings in relation to his position according to the numbers on the clock, as he was taught in his training.

Then, in what at first looked like a mirage reflecting from

the heat of the sun off the stones and water, he saw that there were several police vehicles coming towards him.

"There's our guys," Esposito said, and even the gendarmes straightened up for a glimpse.

The vehicles kept coming. And Nick was expecting an onslaught from the sniper at any moment, but no shots were fired.

Soon they were surrounded by their own men and being loaded into ambulances by the medics. Once inside Nick heard a weak but familiar voice, "Hello, Inspector."

He looked over to the adjacent bed to see a badly injured man. The confusion on Nick's face must have been readable, as the wounded man then croaked, "It's me. Ramon."

Nick laid back down and finally breathed a sigh of relief.

Once Miguel had handed Ramon safely over to the medics, who assured him that the patient was in good hands, he set off for home. Maria and Rosa were overjoyed to see him. They had heard about the crisis at the munitions depot on the radio and had feared the worst. After a brief phone call to Ramon's mother to inform her where her son was being treated, Miguel gave his own mother and sister the whole story, from start to finish.

When he was done, a call to the hospital eased their fears. Despite several broken ribs, a collapsed lung and some internal bleeding due to the repeated beatings, Ramon was expected to eventually recover completely.

Rosa and Maria also had a story to relay. They had had a visitor in the meantime. Professor Kumar had given it one more try to get the computer and notes from Rosa, but she stayed firm. In Maria's emotional state, she had even questioned the professor as to why he would want something that was the cause of such violence. Rosa had explained to Kumar about the attacks, culminating in Ramon's kidnapping, which was news to him. The professor was shocked to hear the news and humbly left with his sincerest apologies.

Professor Kumar left the Rodriguez house feeling like a defeated man. One of his most promising graduate students may be his department's third loss in a matter of weeks.

"Why me?" he thought. He was in a position that he had absolutely no desire to be in. Not only was Ramon, who was going to be one of his newly appointed staff members, in danger, but he was concerned for his own safety. If everyone around him was dying, who knows, maybe he was the next one in line.

The only bright spot of the day came when he returned to his office to find that the strange Mr. Boyd had gone. His secretary said that the visitor had received a call and then quickly left without saying a word. Amar breathed a sigh of relief, retreated into his office, and slumped into his chair completely drained.

With another deep sigh, he switched on his computer, and when the homepage finished loading, he typed two words into the search field 'relaxing vacation'.

Chapter Thirty-Six

Ron and Caroline figured that 8 a.m. local time would not be too early to call Brazil, plus they were so wired and had hardly slept at all, so when their alarm went off at 6 a.m. they were both already awake.

Ron dialed the 12-digit number and the phone rang several times before a sleepy voice answered, "Ola".

"Hi Miguel! It's Ron."

"Oh, hi, Ron."

Ron realized that he had woken his friend and apologized. "I'm sorry Miguel, did I wake you?"

"No, no, it's fine," he said, although he had been in a deep sleep. "Is everything all right?"

"Yes, we are fine. Our flight was great. It's good to be home." But Ron couldn't hold back with more small talk and had to bring up the equation, which was all he could think about. "Miguel, we have the data for the equation. But with all that was going on, I never got Ramon's email address." Ron stifled an embarrassed laugh.

At first there was only silence as reaction and then, "Oh meu deus." Oh, my God. "I forgot about you two."

Nick didn't need a translation for that. "What? What has happened?!?" Caroline then grew tense and gripped his arm.

Miguel let out a deep sigh "Ramon was kidnapped."

"What?!? Oh, no." Ron quickly passed on the information to Caroline then asked, "Is he OK? What happened? I suppose this has to do with FARC and the equation, right?"

"I don't know for sure. But that's our assumption. Anyway, we got him back. He's in the hospital. He was hurt very badly, but he was lucky. They say he will eventually make a full recovery."

"Miguel, please give him our best if you see him, and your family too." Ron didn't know what more he could offer.

"Thanks Ron." Miguel then added, "I'm sorry, but I hadn't even thought about the consequences for you two."

"Of course not," Ron said. "The most important thing is Ramon's recovery and getting him out of the hospital."

"Yes. Yes, I suppose it is," Miguel said. "Look Ron, I'll have to call you back once I find out more about what's going on here."

"Sure, sure. Don't worry about us, Miguel. We'll talk later." Ron tried to sound optimistic, but was deeply disappointed.

Caroline collapsed into the couch. Ron came and sat next to her when he ended the call. Both were in shock and neither spoke for what seemed like an eternity.

Then Caroline said, "I don't know how to feel about this."

"Me neither."

She brushed a tear from her cheek. "I hope Ramon is going to be all right."

"Me, too," was all Ron could say.

"Maybe this is for the best," Caroline added.

Ron looked at her with a confused expression.

"Maybe we won't get to use the equation now after all," she

continued. "It all seemed too easy that way. Almost like cheating."

Ron took his wife in his arms and she began to sob. "It's not like cheating, Caroline. The equation was brought into our lives for a reason, and if it is taken away before we can use it, then there is a reason for that, too. We just have to accept it and move on to our next alternative."

"Which is....?"

Ron didn't reply. There was really nothing he could say in answer.

Caroline reached for a tissue on the coffee table and dried her eyes. "You're right, Ron, but I just don't want to think about it anymore. I just want this all to go away!"

Ron nodded in agreement. "Me, too. But we can't give up. For all we know, Ramon will be brought home and we can move forward just as we'd planned."

Although she didn't really believe it, she replied with a smile, "Yes, I hope that's true." And she went upstairs to shower and dress before she had to wake Sophie for the day.

The police debriefing took place the day after the siege. Nick and Urso had had to spend a couple of hours at the hospital to have their wounds tended to but both had been released to go home and spend the night with their families.

In the meantime the bodies of the gendarmes were removed from the munitions depot and handed over to their comrades.

A dozen or so officers from his team were seated when Nick began summarizing the events of the day. "Congratulations. You all did a great job. Mission accomplished."

Nick paced across the front of the room. "As you all know,

ten FARC members were arrested as they tried to escape via the back exit of the compound. The computer which had been stolen from our facility earlier yesterday was recovered from one of their vehicles."

He then turned to face his officers, "But what some of you might not know is that our hostage has been recovered and is currently safe in protective custody at University Hospital." This led to a rumble of cheers from his team.

He then began to explain that a civilian had found Ramon and brought him to safety.

"Unfortunately, nine gendarmerie members are dead, including the four paratroopers, and the four ground troops who entered the warehouse. Of the four gendarmes who had been stationed outside the warehouse, two were badly injured from the blast, one was killed by a sniper and the fourth was slightly injured by the same sniper. The current status and whereabouts of these three injured parties is unknown."

An officer in the second row commented, "They're probably back at the Vatican by now, laying around in togas getting massages and being hand-fed grapes." A few laughs were heard throughout the room.

But Nick still considered their mission accomplished, since their hostage and equipment were recovered, none of their own officers were lost and only Esposito was slightly wounded.

The officers were then thoroughly briefed about the presence of the gendarmerie and the recent FARC activity in the area. It was FARC which had clearly set the explosives in the adjacent building. They'd simply used it as a decoy to buy them some time to escape if their hideout had been discovered.

Nick continued. "In addition, the university is demanding to have 'their computer' returned, but considering that it is evidence in an ongoing investigation, the computer will be staying put for an indeterminate period. A brief questioning of the hostage did not reveal much information as he was sedated and resting. A complete investigation is pending."

After about an hour he wrapped up the meeting with further instructions. "Referring to the Rodriguez home, we need to keep round-the-clock protection at the residence and at the hospital. The daughter, Maria has been instructed to remain there for the duration of the investigation, for her own safety."

Nick then dismissed the officers and was left seated with only his boss.

Esposito then brought up the most pertinent issue remaining. "The question is how long, Nick? We can't keep watch on them forever. Eventually there has to be some kind of resolution to this, so we can clear out the FARC guys and the Pope's security." He shook his head. "This is so bizarre."

"I've thought about that too." He took a sip of his coffee and leaned back in his chair, which squeaked loudly under his weight. "At the moment, the widow of the professor has possession of the documentation and very likely a functioning equation on the professor's home computer. We believe that Montagne simply copied it onto a USB stick and took it with him when he left there after visiting her that day. Whether the widow has the rights to the information is another issue.

But it seems to me that no one is going to want the rights to this thing when they find out the security issues involved."

Esposito nodded in agreement. "We just experienced that firsthand."

"And so the only way to ensure the safety of Senhora Rodriguez and her family, as well as Ramon, is to stage some kind of public destruction of the information, so that it simply no longer exists."

Esposito let out a humored sigh. He knew Nick all too well and had a very good idea what he had in mind. "Nick, we are not the actor's guild. We are the police force. We don't have the know-how nor the finances to STAGE anything like that."

Nick's sly smile appeared and his boss knew he had something up his sleeve.

"That's right boss, WE can't stage anything, but I know some people, in high places so to speak, who very likely can."

Two days had passed and Ron had not gotten a return call from Miguel. He'd done some searching on the internet and found news about what had gone on in Brazil. A hostage situation with several dead. Ron wanted to call, but Caroline urged him not to. She felt that Miguel's family was much too burdened at the moment and it would be selfish to ask anything of them. And as far as they knew, Ramon may have even died from his wounds. The news reports hadn't listed names, and since almost all the news clips that he could find on the internet were local, they were written in Portuguese, so Ron and Caroline were essentially in the dark.

From her office the next morning, Caroline made the dreaded call to Dr. Blake's office. Chemotherapy would begin the following week.

At almost the same time that Caroline hung up her office phone, Ron was sitting at his computer pressing the "send" button. He knew that at least he had to take the chance. Yes, things were chaotic down in Brazil, and if Caroline didn't want him to call there, he would respect that, but he hadn't promised no contact at all. So he put all the data that Ramon had requested for the Lifespan equation into a file, attached a short letter with well-wishes for his recovery, and sent it to a university email address that he'd found on the internet. His wife and daughter were far too precious to him to give up hope on any possible solutions.

Chapter Thirty-Seven

Maria didn't mind the presence of the security guards, which were assigned to follow her every time she left her mother's home. In fact, she only left the house once a day and that was the time that she looked forward to most. A trip to visit Ramon at the hospital made it worthwhile to get up each morning. Until now, she hadn't spoken to Ramon about the security guards, though she knew that he was aware of them through discussions with the police.

Every time she left the house, she was briefly searched by a female gendarme to make sure that she was not carrying any of the files or paperwork associated with her father's research, then she was allowed to go on her way.

When she entered his room that morning, he was sitting at the table working on his laptop instead of lying in bed. One of his family members must have brought him the laptop and some of his own clothes because he was dressed in a button-down shirt and jeans instead of that horrible hospital nightshirt. He looked like his old self. Her heart skipped a beat as he stood to greet her.

He simply took her hand, kissed it, and asked her to please sit with him. They talked for a few minutes about little things: how she'd slept, how her mother was, who had visited Ramon, and the weather. Then Ramon grew quiet and a somber look crossed his face.

Maria was worried that maybe his feelings for her were not what she had hoped, "Ramon, is there something wrong?"

"I need to speak with you very seriously for a moment, Maria."

Her first impulse was heartbreak, but she was also very curious and nodded for him to continue.

"This work of your fathers. It is putting you in danger."

Maria opened her mouth to object, but Ramon's intense look told her to let him speak.

"I have spoken with the police. They feel that the only way to ensure your safety, as well as that of your family and myself, is to have the work destroyed." He then paused for her reaction.

She dropped her head and very softly said, "I know."

"You do understand why this needs to be done, don't you?"

"Yes."

"I don't want to risk putting you in the situation that I was in. I am lucky to be alive."

Maria was quiet.

"Why are you so sad, Maria?" Ramon asked while again kissing the back of her hand.

When she looked up there was a tear on her cheek. "My father was so wonderful. It breaks my heart to think of destroying something that he created, something that was such an integral part of his life. It was nice to know that some of his work remained. It almost makes me feel as though a part of him is still alive."

They sat for several minutes without speaking. They heard

the clock ticking on the wall, the shuffle of footsteps in the hallway and Maria could only hope that Ramon did not hear the pounding of her heart in her chest as it always did when she was with him.

Ramon finally broke the silence. "You are right, Maria."

She looked up at him, startled by the conviction in his voice.

He sat there nodding, thoughts clearly racing through his head. "Yes, you are right. And you must listen carefully. I am going to tell you what you must do."

A thrill of excitement ran down her spine, and she inched closer to Ramon as he explained what he had in mind.

Chapter Thirty-Eight

"Oh, relax my child. He will call you as soon as he can," Rosa cooed to her daughter.

Maria sat down on the living room couch and let out a sigh of frustration. "I know mother, but I—"

"But nothing. Ramon has been released from the hospital and he is with his family. You must remember that they almost lost him too. If that had happened to you, and I had almost lost you, I wouldn't want to let you out of my sight once I got you back." Rosa came and sat next to Maria. She picked up her daughter's hand and held it tight in her own.

And just then the telephone rang.

Maria was so excited when she hung up the phone that she ran into her room to grab her things, gave her mother a quick kiss and explanation, then ran out the door to her car, only a moment later to return, having forgotten her keys.

Arriving at Ramon's home with a security escort, Maria took a minute to compose herself before ringing the bell, knowing that many of his family members were there. Ramon's mother had invited the aunts, uncles and cousins to supper, since he was feeling so much better and they were all eager to see him themselves. While he was in the hospital his visitations were limited, not so much because of his condition but because of the security issue. Only through his connection with his aunt in the police force was he granted the pleasure of visits from Maria.

This was actually the first time that Maria was being officially presented as his girlfriend to his own family. She should have been nervous, but in fact she was completely at ease once she saw him. When she arrived he took her hand and brought her into the living room where most of the family was assembled. He introduced her to each person, one by one. Most of them had seen her on TV earlier and wanted to hear her description of the spectacle, while others gave condolences on the loss of her father and all that she had been through since then.

A large meal was served whereby she was seated next to Ramon. She felt like the guest of honor. And everyone treated her like she belonged.

When the plates were empty the women began clearing the table and Maria stood to help, but before she had the chance, Ramon got up and began to lead her away.

"No, I must help," Maria said.

Ramon's mother answered. "It's alright Maria. You go with Ramon."

He brought her out into the back garden where they were finally alone.

"Are you okay?" he asked.

She was a bit startled by the question because she actually felt wonderful. "Yes, Ramon, I am having a very nice time."

"That's good. My family likes you."

"How can you know that?" she asked.

He smiled. "I can just tell. Besides… who wouldn't like you?"

Maria blushed and turned away.

Ramon's voice then turned serious as he asked, "Maria, did you have a chance to do what I asked you?"

"Oh, yes, of course! Here… I have it." She reached into her purse and retrieved the USB stick, which she handed to him.

He then gave her a quick kiss on the cheek which caused her to blush again.

"Thank you, Maria."

She simply nodded her head and smiled, happy to please him.

When Maria got home that evening, Inspector da Silva was there with her mother. He talked about the impending situation with them and wanted to be sure that they agreed that destroying the equation was the right thing to do. But by this time, Rosa was not only ready, but eager to get the information out of her hands. It had nearly killed Ramon and had put herself and her entire family in danger. She knew she had made a promise to Caroline, but she couldn't risk her own family for someone she barely knew. She wanted the peace and tranquility of her life back and agreed with Nick on what he believed to be the only way to do that.

Later that evening Maria sat with her mother over a pot of tea and told her everything about her evening with Ramon and his family, dreaming about her future, and all its wonderful possibilities.

At the same time, Ramon sat at his computer and thought of nothing else but the task at hand.

At 2:30 in the morning, he finally finished with his work

and turned off the computer. He looked at his watch and contemplated making the call, but it was late and he was exhausted.

Tomorrow would be interesting, he thought, as he lay down on his bed. Still in his clothes, he was asleep as soon as his head hit the pillow.

Chapter Thirty-Nine

The media spectacle had snow-balled. Everyone in the world wanted to know more about this 'death equation' that had been developed by a professor in Brazil, had been reportedly used twice, and was proven to be accurate. Scientists and mathematicians were interested in its form and function, while theologians wanted to analyze it in order to condemn it properly.

The internet was burdened with searches for "lifespan", "death equation" and "Professor Eduardo Rodriguez". No one knew exactly where the information was being stored or who the rightful property owners were, whether it was the widow of the professor or the university where he worked. Both locations were swarmed with international media. But no one was talking. Neither Rosa nor Maria had made any public statements and, once Miguel was reassured of the safety of his mother and sister, he returned to Brasilia to make sure that his own family was safe. Of course, a small contingent of reporters were camped out there, too.

The media had been alerted by the police that there was to be some kind of an official transfer of the electronic and hand-written data pertaining to the equation. The Rodriguez family would be foregoing their possession of it, but it wasn't clearly stated to whom they would be giving it. The location for the transfer was reported to be the Rodriguez home, as that was where it had been stored all along. The date was set for the following morning.

After an evening of rain storms, which tore down tree branches and took out electricity for most of Sao Paulo, the following morning the sun shone clear and bright, with not a cloud in the sky and not a breath of wind. The atmosphere was almost angelic, perhaps a premonition of what was to come.

The world's attention was focused on Sao Paulo. The media coverage was massive.

And if all went as planned, the subsequent events would be kept in the memories of all who would witness them as surreal and unforgettable. As though in a film, one that was perfectly-orchestrated.

The day began when Father Angelo from St. Francis de Sales church arrived at about seven in the morning. He entered the Rodriguez home, but no further activity was seen for over an hour. At about 8:15, the Sao Paulo police arrived in six official cruisers, accompanied by several unmarked vehicles, which may or may not have been police according to the media. Only two uniformed men entered the house, accompanied by four men in street clothes, one of which was carrying what appeared to be a large suitcase. The rest of the police officers got to work by strengthening up the blockades, cordoning off the premises and immediate vicinity, controlling the media, and working through the crowd with police dogs which were sniffing the entire area, clearly trained to detect explosives.

Another half an hour passed without anything happening. The suspense was building, the media were getting anxious.

At about 9:00 things began to transpire. A large black SUV arrived and out stepped two Cardinals, quickly identified by the press as Cardinal Pena of Brazil and Cardinal Rathborne of the Unites States. As they entered the home, the remaining police officers got out of their cars and took positions in front of the house and on the street. Another vehicle arrived and parked right in front of the house. It looked similar to an armored bank vehicle, one used to transfer cash and valuable goods, but the police logo was plastered on both sides and two officers in heavily armed regalia were in the front seats.

Inside the home, Rosa and Maria sat in front of the television and watched the unfolding events on CNN. They were both amused at the spectacle, but were anxious to have it behind them. Maria was eager to get back to her friends and her job, and Rosa just wanted her peaceful life back.

At ten minutes before ten o'clock the front door of the house opened and the two cardinals walked out. They came forward in grand manner, sweeping across the lawn in their colorful robes, where a makeshift podium was set up with microphones from all the major media outlets. All eyes were on them, from the reporters across the street to millions of people on every continent on the planet watching through live video feed.

The majority of those who had entered the house were now standing behind the Cardinals. There were two men, each holding computers, while another held the large suitcase which was seen earlier being carried inside.

Cardinal Pena approached the microphone. "Good morning," were his first words followed by, "Let us pray." He bowed his head.

Though the next several minutes were followed by prayer, there seemed to be an electricity flowing through the onlookers, all eager to know what the Cardinal would say next.

"As you all know, our dear departed brother, Eduardo Rodriguez, was blessed by God with a brilliant mind," the Cardinal began. "With this gift he was able to produce what most of us would consider a miracle and, in his hands, it was."

Rosa and Maria stood on the front steps with broad smiles, clearly approving of the Cardinal's chosen words.

The Cardinal continued, "But our dear brother has now gone home to be with our Father, and the fate of his work is unclear." Then, referring to Ramon's kidnapping, "As we have already witnessed with near deadly consequences, there are many who would like to acquire this work and use it for their own purposes, some of which are contrary to its original intent." After a brief pause, he added, "This tool is much too powerful for humankind. If it finds its way into the wrong hands, it could be used to do the devil's work." His voice was getting louder with each word. "We cannot let this happen!" He pounded a fist on the podium.

The Cardinal turned around and gestured to Rosa. "Eduardo's beloved widow Rosalinda has agreed to release this work to me, in the name of the Church, so that it may be destroyed."

The crowd gasped and murmurs were heard throughout.

The Cardinal motioned for silence from the onlookers, "Please listen. My colleagues and I, along with the local police, have gathered all known electronic copies of this work. We have thoroughly searched this home and the Professor's office at the university." He motioned to the two men holding computers. "We have Eduardo's home and work computers, which are the only two which he was known to use. We also have the computer of his colleague who also recently demised." He then gestured to the man with the briefcase. "And we have all handwritten copies of notes that accompany the research." He then turned again to Rosa. "Is what I am saying true, Senhora Rodriguez? That I have all copies of the information concerning your husband's work on this topic?"

She nodded and urged herself to speak as loudly and clearly as possible. "You have everything that I know of, Father." And this was true to her knowledge.

"Good. And now..." he paused and turned to his gendarmes behind him, several of which nodded in unison. The crowd was anxious, but his next words were not what anyone expected. "Please follow us," was all he said before he and Cardinal Rathborne climbed into their SUV. The men with the computers as well as the one with the briefcase got into the back of the large armored vehicle, and the remainder of the police dispersed into their cars, with Rosa and Maria getting into the back seat of one of them.

The entire procession of vehicles slowly began down the street.

And the world followed.

Chapter Forty

Ron was the first one up and in the kitchen in the morning. Following routine, he let Max outside first, then began making coffee. He switched on the small TV while waiting for the coffee to brew. The weather forecaster was describing a storm heading their way. Ron looked out the window and saw gray skies and light drizzle and couldn't help but think of the old saying, *April showers bring May flowers.* Always pays to be the optimist.

But then the attention-grabbing music of a special alert on TV brought him back to reality. He sat down to watch what he'd been waiting for. It was a live feed from Sao Paulo. The international television stations had been covering the events around the clock for the past couple of days.

The time had come.

Ron had a dreadful feeling in his stomach.

Once the Cardinal's caravan left, the scene in front of Rosa's house turned to absolute chaos. Reporters were scrambling into vehicles. Some had been lucky enough to have found an early parking spot on the street, while the rest were sprinting to the end of the road where they'd been forced to park behind the barriers due to the roadblock. No one had any idea where the Cardinal was headed, and everyone was on their cell phones trying to get or relay information.

Once the Cardinal was underway, the police leaked the destination to the media. They did not want to risk the chance that what was to occur next would not be recorded and transmitted throughout the world.

The journey from the Rodriguez home to the destination would take nearly forty-five minutes. They would slow it down if necessary to provide plenty of time for everyone to get into place.

It was amazing how fast the media could reposition itself. Within thirty minutes the military zone at Guarulhos Airport was flooded with journalists, mobile communication labs and a sea of satellite dishes.

Cardinal Pena was given running updates. Things were falling into place exactly as planned. He was pleased.

But not everyone was happy. The members of the Board of Regents at the university were outraged. The previous week they had filed a petition in the Sao Paulo court system claiming that they were the rightful owners of all work conducted by their employees, including that of Eduardo Rodriguez, and that his recent work should be returned to the university. On top of that they claimed that the police department as well as Rosalina Rodriguez were in violation of intellectual property rights. Unfortunately for the Regents, the courts were not known for their promptness, and until a court order was extended, they had no power to intervene. Helpless to take action, they too watched the events unfolding on television.

From a hotel suite downtown, the FARC People's Army were considering their options. Once the Cardinal reached his destination, the cameras would be recording everything. That would not be optimal for intervention. The only remaining alternative would be during the transfer, and even that would be difficult, since they had limited technical support. They would have to be creative. A call was placed to headquarters for orders.

The sky was crystal clear and the sun beat down onto the airport tarmac, which began to heat things up, both literally and figuratively. Rumors began to circulate about what was going to happen. Would the Cardinal simply board a plane and fly off with the equation? Perhaps 'destroy' it, as he said, once he returned to Rome? Or was he going to take care of that here before the cameras?

A massive cinema screen stood erected in front of one of the military hangars. There was a projector and a podium. The media were cordoned off several yards away but had been allowed to set up microphones on the podium before being escorted back behind the ropes. Off to the left of the hangar, a UH-60 Black Hawk helicopter was stationed with pilots at the controls, and the engines were idling.

Everyone became even more confused as a cylindrical steel vessel was rolled out of the hanger on a trolley and placed next to the podium. It appeared to be about a meter in diameter and of about the same height. A large coil of chain lay neatly positioned next to it.

Everything was being recorded and reporters were constantly filming updates. Suddenly there were shouts heard among the media. The Cardinal's caravan had been seen entering the airport premises.

Dozens of reporters who had been working in their vehicles now rushed forward to get as close as they could to the action. Cameras were brought into position and microphones were readied. Moments later the black SUV appeared from behind the hangar and was being parked near the podium. The police vehicles were right behind.

Once all the members of the security team were in place, the two Cardinal's and all the minor players assembled themselves around the podium, just as they had at the Rodriguez residence.

Cardinal Pena steadied himself against the podium and looked around. He bowed his head and appeared to be in prayer. The crowd was silenced.

He was now ready to finish playing his role.

The Cardinal then made the sign of the cross and cleared his throat. "We cannot allow the Devil to work among us!" he roared. A few gasps were heard from the onlookers. "We have recognized his attempt to infiltrate our souls with evil!" A spattering of cheers were heard in the crowd.

The Cardinal's look of conviction was most evident for those watching on television, as the strain in his face was visible. No one could doubt his faith.

"Satan has taken the work of a God-fearing man and is now trying to use it against us. We must protect our beloved God!!" More cheers began erupting from those gathered. The press usually stayed neutral as such events, but here they took part like they were supporting members at a rally.

With both hands elevated to the Heavens the Cardinal continued, "The Lord has given it to us and now it is our turn to give back to Him!!!" The crowd was now in a frenzy, some cheering, others jeering and many simply in awe.

The Cardinal then motioned to the computers and notes. "These works will now be destroyed so that we may live according to God's will and not to man's equations!" At those words, the helicopter engines roared to life.

The team behind the Cardinal moved into action. A lid was lifted from the steel vessel and the computers were placed inside. Then the folder of handwritten notes was placed on top. Everything that had been collected earlier from the Rodriguez home and transported to the site was accounted for.

"Is everything inside?" the Cardinal asked.

As a representative of the police force, Nick responded truthfully. "Yes, Cardinal. Everything that we have collected concerning Professor Rodriguez' Lifespan Equation is inside this vessel."

That was the first time that the equation itself had been given an official name, The Lifespan Equation, and the media were furiously taking notes.

The chain was wrapped around the neck of the vessel and the trolley was then rolled over to the waiting helicopter. After securing the long end of the chain to a hitch on the Black Hawk, the trolley was rolled away and the engines of the great machine turned to a deafening high-frequency shrill.

Chapter Forty-One

When Caroline entered the kitchen she immediately knew something was wrong. Ron's face was pale and emotionless. "Sit down", he said, and she took a seat next to him at the table. She expected him to want to talk, but he only pointed to the television and said, "Look."

At first she was completely confused as the clips kept going back and forth between a helicopter and a Cardinal giving a speech, but then a photo of Eduardo Rodriguez was shown and she knew exactly where they were.

Since they had received no further contact from Miguel, she had essentially come to terms with the fact that the equation would not be available for her to use, and she needed to keep her strength up for the battle to come. She wanted to leave what happened in Brazil in her past and begin to move on, but it looked as though that wouldn't be easy. "What's going on?" she asked Ron calmly.

"It's not exactly clear, but the Cardinal was speaking in front of Miguel's mother's house in Sao Paulo." His words broke off as his attention was still glued to the screen.

"And?" she now asked with a touch of annoyance.

"And... they said that they are going to destroy the equation, the work of Miguel's father."

Caroline knew that he was expecting some kind of startled reaction from her, but that didn't happen. She simply stood up and prepared her coffee without a word. Then she shut off the television.

"Why did you do that?" Ron asked.

"Because it does not concern us. We need to move on." She sat down next to him again. "And today I'm going to do just that. I have an appointment with Doctor Blake this afternoon at four o'clock."

Ron put his head in his hands. "But Caroline, you didn't talk to me about this."

"I know. I'm sorry, I should have, but, well, I just thought it was the only option left, and until now I've just wanted to block it out of my mind and live normally."

"We do live normally. And we will," Ron concurred. "But we need to communicate with each other. I know I've also isolated recently, but as of now that is going to change. We need to put the past behind us and move forward together, okay?"

Caroline got up and went to her husband. He pushed his chair away from the table and she settled onto his lap. They enjoyed a brief moment of silence before Sophie came tearing into the kitchen clamoring for breakfast.

The room fell silent once again as soon as Sophie was given her bowl of Corn Flakes.

Caroline stared off into the distance for a moment and found herself deep in thought, "It's funny."

"What's funny?" Ron asked.

"I don't know if I can explain it, but I just have such a strong feeling that I'm going to be OK."

"I'm glad you feel that way, but...," he hesitated, "... well, let's just take one step at a time. I'm sure you are right... the chemotherapy is the right thing to do."

She let out a small laugh. "Actually I don't know if it has to do with the chemotherapy at all, more like a..., well, a twist of fate. Something is waiting in the wings."

"What does that mean, mommy?" Sophie asked, "Waiting in the wings? Like sitting in a bird's feathers?"

Caroline smiled. "That's exactly right, Sophie. I feel like I'm sitting among the feathers of a giant bird and she is taking care of me."

Sophie giggled.

Ron added dubiously, "I'm not sure about fate, Caroline, nor much about birds, but whatever the reason is for your mood, I'm thankful for it."

Chapter Forty-Two

Miguel woke up feeling terrible. He knew the day that lay ahead of him would be trying, to say the least. His mother had called the night before and explained everything about what was to occur the next day, and although he had mixed feelings about not being there with his sister and mother, he knew that he needed to be near his wife and children.

On top of that he was feeling guilty about not having been in contact with Ron since his one brief phone call the week before. His father's work would not be available for Caroline to use, and although he believed the work had to be destroyed in order to keep his family safe, he couldn't help but somehow feel responsible for Caroline's fate. He knew he had a telephone call to make. And it had to be today.

There were two police officers posted outside his house for security measures, but even so, he wanted to be with his family that morning while his father's work was being destroyed. He did not turn on the television for fear of upsetting his pregnant wife, so they lingered over breakfast until he finally had to get to the office.

Miguel was giving Katie a kiss goodbye at the door when the phone rang. She ran to answer it as Miguel walked out to his car. Then he saw Katie standing in the doorway with the phone held out to him. "Miguel! É Ramon!"

He got out of his car, went back into the house and took the phone from his wife. "Ramon. This is a surprise. How are you?"

Ramon simply responded that he was doing very well, but was eager to get to the point of the call. As Ramon began explaining, Miguel drew a deep breath and edged over to the back of the couch for support.

Katie saw the look on Miguel's face and asked, "Está tudo bem?" Is everything alright?

Miguel nodded. He was speechless. After a few affirmative remarks he hung up the phone. Katie came to him, looking for answers. At first he had tears in his eyes.

And then all he could do was laugh.

"Mama?"

"Yes, Sophie?"

"Which is colder? Ice or snow?"

Ron let out a small laugh and looked at his wife, amusingly waiting for her answer.

"Well, that depends." She drew out the last word and Ron knew that his wife was trying to buy time to come up with a good response.

"Depends on what, mommy?"

"Ok. Imagine you put some snow into the freezer right next to the ice cubes."

Sophie's eyes grew wide as she was surely trying to picture some fluffy white snow next to the ice cube tray.

"Now, which do *you* think is colder?" Caroline said.

"I don't know mommy, that's why I asked you!"

Ron gave another laugh. "Smart kid."

"Okay, Sophie. Think of it this way, if the temperature in the freezer is say, 25 degrees, then everything in there is also 25 degrees. The frozen pizza, the ice cream, even the snow and ice cubes."

Sophie was thinking for several seconds, then said, "Oh! I know, they are the same!"

"Right. And if the snow and ice were outside, and the temperature outside was colder than in the freezer, then they would both be colder than in the freezer."

Sophie nodded.

Caroline continued and was speaking very slowly so that her daughter could understand. "BUT if the snow was outside and the ice were in the freezer, and it was *colder* outside.... then which would be colder?"

She got it right away. "The snow!"

"That's right. The snow." Caroline gave her daughter a kiss on the nose. "Your daddy is right, too. You are a smart kid."

Sophie beamed.

Ron was thinking that this was exactly the kind of normalcy that he needed to try to keep in his family life. Thank goodness for Sophie. She makes it all easier to keep perspective on what's important.

The wonderful family moment was suddenly interrupted by the doorbell. It was Sophie's ride. A neighbor who also had a child in the same daycare had offered to carpool, so she picked up Sophie every other morning. Caroline helped Sophie with her shoes and gave her a kiss goodbye. Just as she closed the door, Ron's phone rang.

"Who could that be?" Caroline asked. "I hope its good news." She laughed with a note of sarcasm.

Ron picked up his phone and was surprised to hear Miguel's voice. "Miguel. Hello."

That caught Caroline's attention and she came to his side, catching the first few formalities in greeting.

"Ron, is Caroline there, too?"

Ron was not sure why he was getting such an early morning phone call, nor why he was asking for Caroline. A bit hesitantly he replied, "Yes, she is here. Would you like to speak with her?"

"No, actually I wanted you both to hear what I have to say. Can you put me on speakerphone?"

Ron switched to speaker phone. "Okay, Miguel, we can both hear you now."

"Good morning, Miguel." Caroline said.

"Hello, Caroline. How are you?" Miguel asked.

"I'm doing fine. How are you?"

"Very well, thanks." He paused. And Caroline and Ron exchanged confused looks.

"Caroline. Ron. I just got an interesting phone call from Ramon."

Caroline thought that his injuries were worse than she had heard, or maybe his life had been threatened again. "Has something happened?"

Miguel reassured them. "No, no. Everything is fine. Actually, more than fine. Listen. What I am about to tell you must remain between the two of you... for the safety of my family."

Ron and Caroline answered in unison, "Of course."

"Ramon has an answer for you."

Caroline was utterly confused, but Ron knew exactly what Miguel was referring to.

"Oh my God," was all Ron could say as he collapsed against the couch.

"What?" asked Caroline beginning to get worried.

"Ron?" Miguel wondered if he should continue.

"Just a moment Miguel," Ron answered and looked to his wife. "Honey, I sent your data to Ramon."

She too, then, knew exactly what he meant, and she reached to the wall for support. "But I thought the equation was destroyed?"

Miguel answered. "I thought so too. I mean, it is to be destroyed or has been by now for all I know. But apparently my sister and Ramon... well, apparently, the two of them were in on this together." His voice trailed off before adding a near-forgotten formality. "Oh, Ramon said he would have contacted you directly, but he said that his English is so poor."

"Okay, Miguel, that doesn't matter," Ron interjected, barely able to wait any longer. "Does Ramon really have an answer for us?"

Tears began to stream down Caroline's cheeks as she was overtaken with emotion. She began to take short breaths. "Oh, no. Oh, no," she repeated over and over.

Ron put his arms around his wife. "Shhhhhh. It's going to be alright." Ron could feel her body shaking, and wondered himself if he had done the right thing.

Miguel's voice was again heard. "Ron? Caroline? May I continue?"

Ron looked to his wife. "Are we sure we want to know?"

She took a deep breath, then nodded.

Chapter Forty-Three

The noise was deafening as the Black Hawk lifted off the ground. The chain also began to rise and soon the vessel was airborne as well. It took off in a south-easterly direction. The media wasn't sure what to do or where to go. How would they know what was to happen?

After only moments a second helicopter was seen joining the Black Hawk. It didn't appear to be military, but rather a news helicopter. And then a live video feed was displayed on the movie screen positioned in front of the hangar. Clearly the second helicopter was filming the Black Hawk. The media were elated and cheers erupted from the onlookers.

Both Cardinals returned to their vehicles, but the police as well as Rosa and Maria remained outside, all eyes glued to the screen.

After several minutes of simply watching the bird in flight, the media got anxious and began shouting questions to the police officers.

"Now what? Is it over?"

"Where are they going?"

"Do you know the destination so we can meet them? Or send other teams?"

"Who is paying for this?"

Nick just shook his head and shrugged his shoulders in the universal gesture of obliviousness.

Gasps were heard from the crowd as the ocean came into view. Were they taking it out to sea? Some even speculated if a Black Hawk could make it as far as Rome. But then why the vessel? If the Cardinals intended to take it with them to the Vatican, then why not just pack up the hard drives and fly home?

It then appeared as though the Black Hawk was pulling away from the smaller helicopter. The cameras were zooming in on their target, but there was clearly an increasing distance between the two aircraft.

The Black Hawk was now over the water and the distance from shore was being speculated. A mile, now two?

The FARC leaders were frustrated that they hadn't been able to intercept the Cardinal's caravan. But now, watching the unfolding drama on television, they began to see some hope on the horizon. Were the gendarmes simply going to drop the data and computers into the ocean? Could it really be that simple? Depending on how far from shore they were, they could easily get a team out there and search for it. Finding it would be another story, but at least all hope was not lost.

Then it happened.

The Black Hawk ascended slightly, released the chain and vessel, and banked hard right one-hundred eighty degrees. The helicopter was now headed rapidly back to shore as the vessel plummeted toward the dark water.

The vessel made a small splash, then all traces were washed away as the next wave passed.

The crowd was silenced.

And then, just as everyone expected the camera footage to end, an enormous plume of water erupted from the sea from what must have been a massive underwater explosion.

The vessel had been full of explosives. And detonated.

And at that moment it was clear, to the world as its witness, that the Lifespan Equation was no more.

Chapter Forty-Four

"Go ahead, Miguel," replied Ron, "we are listening."

"I am surprised as you are about this. Like I said, I got a call from Ramon just moments ago," Miguel paused. "I don't have an exact date for you though."

Ron let out a sigh of disappointment, or perhaps relief, he wasn't sure. "Didn't it work? Wasn't the data accurate enough?"

"Oh, yes I believe it was. And Ramon has certainly gotten an answer, but he refused to give it to me."

Caroline thought she understood. The date must lie so close in the future that he didn't want to curse her last bit of time on earth. In almost a whisper she said, "We understand Miguel, but thanks anyway."

"No, no, you misunderstand. He refused to give me the exact date because he said that it was irrelevant. Let me think, in English....what he actually said was that, 'you may still be around to see your grandchildren get married.'"

Ron thought he must have heard wrong. It just didn't register. Neither Caroline nor Ron said a word. They just stared blankly at the phone.

"Ron? Caroline? Are you still there?"

Ron grabbed hold of his wife and the tears of joy began to flow.

"Miguel, you are a god-send!" Caroline shouted.

Ron was just speechless and vaguely registered Miguel's voice in the background saying how happy he was for them.

After about a thousand thank-yous and a promise to talk soon, Ron and Caroline said goodbye to Miguel.

Caroline walked over to where her purse was sitting on the small table near the door. She pulled out a business card and came back over to Ron on the couch.

"What's that?" he asked.

She ripped the business card into shreds. "That *was* my appointment card for the chemotherapy. I won't be needing it anymore."

Ron smiled. Then he scooped his wife up off her feet and carried her upstairs.

It was time to start on that second child.

Epilogue

Later that day in Vatican City...

Deep in the catacombs beneath St. Peter's Basilica, an ageing priest shuffled along a stone corridor, the only sound was his leather sandals scraping against the centuries-old limestone.

He carried a small, hermetically-sealed box.

Place this in storage, he was told by the Cardinal.

He was given no further instructions. So he chose a simple enclave near the tomb of St. Ignatius of Antioch.

And there, he left the works of a Brazilian mathematician, in peace and solitude, for eternity.

...Or so it would seem.

About the Author

Holly Zimmermann was born in Providence, Rhode Island, USA with the maiden name Morrison, and was the middle child of three girls. She has degrees in Mechanical Engineering from Worcester Polytechnic Institute and The University of Rhode Island as well as an MBA from The College of William & Mary in Virginia. She worked for many years in R&D for military applications including antenna and radar jamming systems for tactical fighter aircraft and later acoustic damping for submarines.

Apart from writing, she spends her free time running and taking part in adventure sports around the world including the 257-kilometer Marathon des Sables race across the Sahara Desert and the Polar Circle Marathon in Greenland. In 2018, she was the international women's winner of the Mount Everest Marathon, the highest marathon in the world. She is a regular public speaker at companies, schools and sporting events.

Her book, *Ultramarathon Mom: From the Sahara to the Arctic*, was released in April 2018, and *Running Everest: Adventures at the Top of the World* in April 2020, both published by Meyer & Meyer Sport.

Holly can be found online at:

Website: www.hollyzimmermann.com
Instagram: holly.zimmermann
Facebook: @ultramarathonmom;
   or www.facebook.com/holly.zimmermann1
Twitter: HollyZimmermann@gehmalaufen
   (*gehmalaufen* means 'Let's go running' in Bavarian)
YouTube: www.youtube.com/c/HollyZimmermannNeverGiveUP
Snapchat: hzimmermann27

Contact Holly via email at: hollyzimemrmann@yahoo.com

www.ingramcontent.com/pod-product-compliance
Lightning Source LLC
Chambersburg PA
CBHW071803080526
44589CB00012B/661